To the memory of my father, Cy Waldman, a gentle man.
– MW

To my daughter Keri and all my nieces and nephews.
The future is yours.
– MJL

Dying for a Hamburger

The alarming link between the
meat industry and Alzheimer's disease

Dr Murray Waldman
and Marjorie Lamb

PIATKUS

Copyright © 2004 by Dr Murray Waldman and Marjorie Lamb

First published in Canada in 2004 by
McLelland & Stewart Ltd

First published in Great Britain in 2004 by
Piatkus Books Ltd
5 Windmill Street
London W1T 2JA
e-mail: info@piatkus.co.uk

The moral right of the author has been assserted

A catalouge record for this book is
available from the British Library

ISBN 0 7499 2554 X

The Standardized Mini Mental State Examination (SMMSE) on pages 30–32 is reprinted by permission of Dr William Malloy.

Printed and bound in Denmark by
AIT Nørhaven A/S, Viborg

Dying
for a
Hamburger

Contents

Dying
for a
Hamburger

PREFACE

For thousands of years the elderly were regarded as wise. It's only in the last few decades that this attitude has undergone a change. Now the elderly are often regarded as foolish and forgetful people who must be cared for as if they were children. The reason for this transformation is simple: Almost half of those older than eighty-five are suffering from Alzheimer's disease (AD).

During the research for this book several startling but easily verifiable facts came to light.

First, Alzheimer's is a relatively new disease. Prior to about 1900 there is no mention of this disease in any medical, religious, or secular literature. We find dementia associated with syphilis or vascular disease, but nothing resembling AD. Alzheimer's did not even have a name.

Secondly, although Alzheimer's is a disease of the elderly, the reason for our current epidemic of AD is not related to our increased lifespan. Most of the gains in life expectancy have been made at the bottom of the scale. That is, infant mortality has declined sharply in the past century. But those individuals, hundreds of years ago, who survived to middle age could look forward with some confidence to a relatively healthy old age. Even before the advances made in medicine and hygiene in the twentieth century, thousands of people lived well into their seventies, eighties, and nineties – and yet we find no evidence of an epidemic of dementia in that population corresponding to the numbers we see today.

Thirdly, Alzheimer's appears to have emerged in different parts of the world at different times over the last one hundred years, and there are still large parts of the world where it is relatively unknown. In these regions, Creutzfeldt-Jakob disease (CJD) is also rare. We will examine evidence that variations in diet may explain why this is so. We'll look at a relatively unknown infectious agent called a prion, consider the horrible neurological diseases that it causes, and see how prions could become widely distributed in our food. The history of the mad cow, or BSE, epidemic in Great Britain may provide a clue.

The rapid rise in the rate of prion diseases is one of the most fascinating stories in modern medicine. It is a story of cannibals, both two- and four-legged, of men and women who ended up dying in dreadful ways because they wished to grow taller or have children, and of inadvertent government and industrial bungling, some of which continues to this day.

This is a cautionary tale of what happens when science lags behind industrial innovation, and how nature can sometimes wreak terrible vengeance on those who are completely unaware that they have violated her laws.

Remarkably it's also a story with a happy ending. Instead of working toward simply alleviating the symptoms of neurological diseases or looking for a cure, we can work toward prevention. Prion diseases have a long latency period, so the current rates of disease probably reflect practices that were current from ten to twenty years ago – practices that conformed to generally accepted government and scientific norms of the time. It may turn out to be the case that these practices, as presently organized, need to be changed for health reasons. But industry cannot reasonably be blamed for not knowing this in advance of the scientists. And changes are already being implemented in the way meat is processed and tested in most countries of the world. Perhaps with a few other simple measures, twenty years from now AD will be on its way to becoming an historical footnote, comparable to the Black Death.

ACRONYMS

ADC	AIDS dementia complex
AHAF	American Health Assistance Foundation
AHPHP	Australian Human Pituitary Hormone Program
AIDS	acquired immune deficiency syndrome
ALS	amyotrophic lateral sclerosis (i.e., Lou Gehrig's disease)
APHIS	Animal and Plant Health Inspection Service (U.S. Department of Agriculture)
BSE	bovine spongiform encephalopathy (i.e., mad cow disease)
CDC	Centers for Disease Control and Prevention (U.S.A.)
CFIA	Canadian Food Inspection Agency
CJD	Creutzfeldt-Jakob disease
CJDSU	Creutzfeldt-Jakob disease Surveillance Unit (U.K.)
CT	computerized tomography
CVL	Central Veterinary Laboratory (U.K.)
CWD	chronic wasting disease (deer and elk)
DLB	dementia with Lewy bodies
DNA	deoxyribonucleic acid
ECG	electrocardiogram
EEG	electroencephalogram
EU	European Union
(f)CJD	familial CJD
FDA	Food and Drug Administration (U.S.A.)
FFI	fatal familial insomnia
FSE	feline spongiform encephalopathy
FSH	follicle stimulating hormone
GHD	growth hormone deficiency
GSS	Gerstmann-Sträussler-Scheinker disease
HAART	highly active antiretroviral therapy
HGF	Human Growth Foundation (U.S.)
hGH	human growth hormone
HIV	human immunodeficiency virus
hPG	human pituitary gonadotrophin
LH	luteinizing hormone

3

MAFF   Ministry of Agriculture, Fisheries, and Food (U.K.)
MBM   meat and bone meal
MID   multi-infarct dementia (i.e., vascular dementia, VaD)
MMSE   Folstein Mini-Mental State Examination
MRC   Medical Research Council (U.K.)
MRI   magnetic resonance imaging
MRI   magnetic resonance imaging
MRM   mechanically recovered meat
MSM   men who have sex with men
NIH   National Institutes of Health (U.S.A.)
NPA   National Pituitary Agency (U.S.A.)
NPRP   National Prion Research Program
NPU   Neuropathogenesis Unit (Edinburgh)
OIE   Office International des Epizooties
PCRM   Physicians Committee for Responsible Medicine
PPS   pentosan polysulphate
RNA   ribonucleic acid
SARS   severe acute respiratory syndrome
SBO   specified bovine offal
SEAC   Spongiform Encephalopathy Advisory Committee (U.K.)
SMMSE   Standardized Mini-Mental State Examination
SRM   specified risk material
TIA   transient ischemic attack
TME   transmissible mink encephalopathy
TSE   transmissible spongiform encephalopathy
USDA   United States Department of Agriculture
VCJD   variant Creutzfeldt-Jakob disease (i.e., human BSE)
WHO   World Health Organization

PART ONE

The Short History
of Alzheimer's Disease

CHAPTER 1

DEMENTIA: HERE AND THERE

When he was about fifty years old, my father collapsed over dinner at a restaurant one evening. My father was a quiet man. He enjoyed reading, playing poker and bridge, swimming, and playing golf. He managed a mid-size business, and except for some high blood pressure and mild diabetes, which he controlled by diet, he was never really sick.

As luck would have it, on the night he collapsed, he and my mother were dining with a couple who were both physicians. My father was whisked to the hospital, but by the time he arrived he was already feeling better. Nevertheless, he underwent a battery of tests – none of which showed anything abnormal. The diagnosis was "syncope NYD," which means "fainting, not yet diagnosed." The next day he was released and went home.

Following this event, my father seemed completely unchanged physically, but he never read books again, saying he had difficulty concentrating on them for long periods. He still read magazines, and began to watch TV more.

Over the next ten years he had many more small events. Most went completely unnoticed at the time, but his speech might be slurred a bit for a few days, or he might wake up and find a few fingers had gone numb. Soon he had stopped swimming, saying that it didn't interest him any more. Once, for a period of a few weeks he walked tilted to the right.

When asked why he was walking in this strange fashion, he said he was unaware that anything was wrong.

His hands became clumsy, and he had difficulty cutting his food or holding cards. He stopped playing poker with his buddies on Sunday afternoons. Later he stopped playing bridge with my mother and their friends. He retired from work and began to spend his days watching TV. If asked to do some therapy to increase his strength or dexterity, he would agree for a few minutes and then lose interest. His room was filled with balls and other paraphernalia that his children, on advice from his doctors, bought him in order to help him recover his fine motor skills.

He became mildly confused, lost track of the date, forgot what he had done in the previous few days. My brother took over the management of his financial affairs. As time went on, my father had more and more difficulty walking, until eventually he was confined to his chair. He became incontinent, requiring a catheter and diapers.

Never during this ordeal – which lasted well over twenty years – did he ever complain either of pain or about his increasing disability. Unlike Alzheimer's disease victims, who experience a relatively rapid slide into dementia, my father seemed to drift downward at a measured pace. It was as if he were slowly descending a gigantic staircase. For weeks or months his condition would remain the same, then overnight he would lose another function, or suddenly a mental task he could perform the day before would be beyond him. Throughout the entire experience he remained calm, undemanding, and uncomplaining.

Finally he succumbed very quickly to an overwhelming infection that resulted from his having to use a urinary catheter. His death was undoubtedly hastened by his generalized weakness due to his years of immobility.

As is always the outcome of progressive dementia, he eventually just stopped breathing. But like so many other people who have lost a parent to dementia, we all felt that tiny parts of him had died long before his final demise.

THE SCOPE OF DEMENTIA: FOUR TERRIBLE PLAGUES

Derangement, madness, lunacy, craziness, delusion, mania, psychosis – these are some of the terms people use when they talk about dementia.

Professor W.A. Lishman, emeritus professor of neuropsychiatry at the Institute of Psychiatry, University of London, defines dementia as "an acquired global impairment of intellect, memory, and personality, but without impairment of consciousness."[1] Note that dementia is acquired – in other words, one is not born demented, as one might be born mentally retarded. Dementia is a decline that happens to a person who once functioned on a higher level.

Further signs of dementia include deficits in several areas of cognitive function, for example, loss of memory, loss of ability to solve simple puzzles, and loss of social inhibitions. Dementia is, for the most part, a gradual and irreversible process. (A very few causes of dementia are treatable. These include normal pressure hydrocephalus, brain tumours, and dementia due to metabolic causes and infections.) In writings before about 1800, almost all dementias were blamed on poor circulation to the brain or bad habits that the sufferer indulged in. *A Treatise on Diseases of the Nervous System*, published in 1871, describes the causes of dementia, as they were then understood:

> Among the physical causes, drunkenness, the use of opium, and other narcotics, excessive venereal indulgence, masturbation, blows on the head, exposure to severe heat or cold, the puerperal state [being pregnant], and certain diseases may be referred to.[2]

The description implies a moral judgment upon victims of certain dementias, a contempt that continues in many quarters today, and in many cases delays prevention or treatment.

Dementia, as presently understood, takes four principal forms, varying according to cause. Of the four, only multi-infarct dementia existed in the past.

- Multi-infarct dementia (MID) is, in part, a result of lifestyle.
- HIV/AIDS dementia is caused by a virus.
- Dementia with Lewy bodies (DLB), although common, is difficult to diagnose.
- Alzheimer's disease (AD), generally said to be of unknown

cause, may be due to another infectious agent, a prion,
that has only recently been identified.

Multi-Infarct Dementia

Descriptions of multi-infarct dementia (MID) cases can be found readily
in medical literature. One of the best early descriptions of this condition
is recorded in a book called *Clinical Lectures on Mental Disease* by T.S.
Clouston, published in 1898. In his lecture, Clouston describes the case
of Mr. L.D., age seventy-eight.

> [Mr. L.D.] had been hard working and as drunken as his lim-
> ited means would allow. Senile insanity is often the penalty
> for an excessive use of alcohol earlier in life. About nine
> months ago he got a fall down stairs and has not been so strong
> or well since. About six months ago his memory began to fail,
> then he became stupid and confused, then suspicious, then
> restless, then unmanageable, then violent to his wife, and was
> then sent to the Asylum as a pauper patient.
>
> On admission he was confused, slightly excited, very rest-
> less, his memory gone, his general condition weak, his senses
> blunted, his speech senile, his pupils irregular in outline, his
> tongue tremulous, his pulse 90, weak and intermittent, his
> temperature 98.2, his lungs and other organs healthy, and
> his appetite good. He was well fed and nursed in our hospital
> ward, but though he gained in flesh he did not improve. He
> was restless, especially at night, became gradually dirty in his
> habits, moved about in a purposeless way all the time. The
> motor restlessness of a senile case is an extraordinarily vital
> phenomenon. He never sits down, seldom sleeps, he shouts,
> and walks about his room all night, yet never tires.[3]

The author describes how the man begins falling down. He breaks
his hip and is then confined to bed. Still very restless and agitated, he
becomes incontinent of urine and stool, and eventually dies. An
autopsy is done, and the report describes evidence of arteriosclerosis (or

hardening of the arteries) in the cerebral blood vessels. Physicians of that era assumed that as these vessels became narrowed, the decrease in blood flow to the brain led to the patient's progressive mental decline. Today we know that arteriosclerosis is a symptom of a generalized disease process.

Vascular disease is the result of fatty deposits in the blood vessels, known as plaque. In response to various stressors, these plaques can break loose from a vessel or from the interior of the heart and travel through the bloodstream to the brain, where they form a plug in an already narrowed blood vessel.

"Downstream" from the blockage, this plug deprives the brain of oxygen, resulting in an area where the cells die of hypoxia, or lack of oxygen, causing an infarct. Infarct refers to a localized area of dead tissue caused by inadequate blood supply. The common term for an infarct in the brain is a stroke. "Multi-infarct" means that the accumulation of blockages to the small blood vessels has killed or damaged cells in numerous areas. Multi-infarct dementia is most commonly described as being the result of a series of small strokes.

Strokes can range from the extremely mild, resulting in perhaps only a few seconds of numbness or dizziness, to major events resulting in coma or death. Each year about 5 million people die of stroke and 15 million have non-fatal strokes. There are more than 50 million survivors of stroke alive, worldwide, today. One in five stroke survivors will have another stroke within five years.[4] Many of those stroke survivors will develop dementia as a result of their multiple strokes.

About 10 to 20 per cent of all dementias are caused by the narrowing blood vessels, making MID the third most common cause of dementia in the elderly.

MID affects men more often than women and usually affects people older than fifty-five years of age, with the onset averaging around age sixty-five.

The major risk factors for MID are divided into two: those that are treatable, such as high blood pressure, and those that cannot be treated or in which the treatment does not seem to affect the course of the disease.

My father, who fell victim to MID, had several of the risk factors: he'd been a smoker for many years (quitting in his late forties), he had a bad family history – his father died of a stroke and his mother had high blood pressure and heart disease. He inherited the high blood pressure, and in addition was in later life a mild diabetic.

Risk factors for vascular dementia include all the risk factors for stroke. The treatable group of risk factors includes:[5]

- high blood pressure;
- heart disease;
- small, reversible strokes called TIAs (transient ischemic attacks);
- overly thick blood (polycythemia rubra vera);
- sickle cell disease; and
- cigarette smoking.

In other words, if you suffer from any of the above maladies, treating the condition will lessen your risk of heart attack or stroke, and therefore your risk of vascular dementia.

Risk factors that are untreatable or whose treatment is not of established value include:

- age;
- gender (men and women seem to get it at about the same rate, but more women than men die of stroke);
- family history;
- ethnicity;
- diabetes mellitus;
- previous stroke; and
- partial blockage of the main arteries supplying the brain (the carotid arteries).

A third group of factors seems to increase the odds of having a stroke, but hard statistical evidence is lacking. These include:

- season and climate (stroke deaths occur more frequently when it's very hot or very cold);

- socioeconomic factors (poorer people seem at greater risk than wealthier ones);
- excessive alcohol intake (more than two drinks a day for men and binge drinking); and
- cocaine or other intravenous drug abuse.

Amongst those risk factors, alcohol seems to generate the most controversy. Long before St. Paul exhorted Timothy to "use a little wine for thy stomach's sake and thine often infirmities," enlightened societies appreciated the benefits of drinking in moderation. Today, medical research continues to probe the French Paradox: Why do French men who smoke, are physically inactive, and enjoy a cholesterol-laden diet have a relatively low rate of heart attack and stroke? Along with their theoretically deadly diet, the French consume more wine – and more alcohol of any kind – than any other country in the world.

We now know that wine, or more accurately the alcohol in the wine, provides at least one protective factor. While high alcohol consumption raises one's risk of heart attack and stroke, moderate consumption actually lowers the risk. So it's not the teetotaller whose lifestyle is the healthiest, but the individual who enjoys a glass or two of claret with his or her evening meal.[6]

HIV/AIDS and Dementia

Acquired Immune Deficiency Syndrome (AIDS) is recognized as one of the deadliest plagues of the modern era, and it has by no means run its course. Dementia is one side effect of the disease experienced by a proportion of AIDS victims. HIV/AIDS differs from MID, dementia with Lewy bodies, and Alzheimer's disease for two reasons. First, the cause of the disease, how it spreads, and methods for prevention are all well understood – and there is even excellent progress being made in the area of treatment. Furthermore, vaccines hold out some promise for long-term prevention and eventual eradication of the disease. The second feature that sets HIV/AIDS apart from most of the other dementias is the relative youth of its victims.

Although AIDS dementia complex (ADC) is only one symptom of AIDS, it occurs in 30 to 60 per cent of untreated sufferers, and represents a huge additional burden on those caring for people suffering from this disease.

Early symptoms include memory loss, problems with concentration and attention, declining strength, dexterity, and coordination, and behavioural changes. Later, victims cannot move or speak properly, and enter a chronic state of altered consciousness, in which the victim appears alert intermittently, but is not responsive. A patient in the end stage of ADC is in a nearly vegetative state, exhibiting only rudimentary intellectual and social comprehension and output. The patient becomes nearly or absolutely mute, exhibits weakness or paralysis in the lower limbs, and has urinary and fecal incontinence. As with other dementias, ADC ends in death – often within six months.

Prior to effective antiretroviral therapy, ADC occurred in more than 60 per cent of patients who developed AIDS. However, with the use of highly active antiretroviral therapy (HAART) – where it is available – the incidence and prevalence of ADC have declined sharply.

In the United States, of patients with HIV/AIDS on HAART, 20 per cent develop ADC. HAART not only has improved the prognosis of AIDS in general but also may reduce the incidence of ADC, delay its onset, and lead to improvement in cognitive function in patients who already have ADC. Regrettably, as with many other therapies, HAART is not in wide-scale use in economically disadvantaged areas of the world, such as Africa, where it is most needed.

Dementia with Lewy Bodies

A third cause of dementia in the elderly, although it cannot be described as an epidemic, is dementia with Lewy bodies (DLB). Lewy bodies are small inclusions (or lumps) found in vacuoles of injured or fragmented neurons. In 1914, Frederick Lewy first described these dense, plaquelike bodies found in the brain cells of people with this disease. Lewy bodies are usually circular with a dense protein core surrounded by a halo.

Dementia with Lewy bodies was first reported in a series of cases from Japan in the 1970s. Once believed to be a rare condition, DLB is now recognized as the second most common form of dementia after Alzheimer's disease, although data on incidence and prevalence are scarce. Patients with the disease have symptoms similar to those of Alzheimer's and Parkinson's patients, making DLB difficult to diagnose. Many investigators believe that a spectrum of Lewy-body disorders exists.

Autopsy studies in the United States, Europe, and Japan suggest that DLB accounts for 10 to 20 per cent of dementias. Up to 40 per cent of patients with Alzheimer's disease (AD) have concomitant Lewy bodies.

Like AD, DLB is a disease of late middle age and old age. Symptoms of DLB are similar to those of AD:

- progressive dementia;
- impaired mobility and balance; and
- loss of control of bodily functions.

However, with DLB early symptoms often include complex and detailed visual hallucinations and difficulties with movement similar to Parkinson's disease. This has led some researchers to conclude that it is something like a combination of AD and Parkinson's. In the initial stages, patients experience extreme fluctuations in cognitive performance (attentive and talkative one day, drowsy and mute the next).

As the disease progresses, DLB results in a profound dementia, with death ensuing in about seven years from the onset of symptoms – approximately the same time frame as "classical" AD.

Because of its neurochemical differences, the cognitive impairment of DLB may be more amenable to drug therapy than Alzheimer's disease.

Although the clinical picture is quite striking, including hallucinations and difficulty with walking, it wasn't described until the 1970s. The brains of the victims have a unique appearance when viewed under the microscope; however, these descriptions also did not appear in the literature until well into the second half of the twentieth century.[7]

Alzheimer's Disease

Alzheimer's disease has become, over the course of the past half-century or so, the most widely recognized form of dementia in old people. Its causes and effects have been the subject of a great deal of speculation and research in recent years. The number of its victims is growing. Although this number has not yet approached the total killed in the great plagues of centuries past, AD is already epidemic in proportion. Even as the epidemic of AIDS begins to wane in developed countries, the projection for AD victims continues to rise into the many millions.

Ironically, some of the countries with the greatest AIDS problems have not yet suffered the epidemic of Alzheimer's disease currently underway in North America and western Europe. India has the lowest rate of Alzheimer's disease in the world, followed closely by equatorial Africa. The rates in China, although rising rapidly, are still much lower than in Europe and the Americas. Russia appears to have fared the worst, having high rates of HIV/AIDS, Alzheimer's, and vascular dementia. This triple burden upon the nation may play a role in Russia's economic difficulties.

DEMENTIA IN THE PACIFIC RIM

Until about 1970, vascular disease or multi-infarct dementia was the principal cause of dementia in Pacific Rim countries, notably China, Korea, and Japan. (The same was probably true for other countries in that area, but poor public-health statistics preclude definitive pronouncements on the topic.)

There are two reasons why this form of dementia was so common in these areas – cigarette smoking and high blood pressure.

About 1.1 billion people regularly smoke cigarettes – roughly one-third of the global population aged fifteen years and older. The Pacific Rim has the highest rate of cigarette smoking in the world. In 2000, within the Western Pacific Region, which covers East Asia and the Pacific, there were 434 million smokers, of whom almost 400 million are men. This amounts to the highest prevalence of male smoking among all regions in the world – 62 per cent for men, while only 6 per cent for women.

One of every three cigarettes consumed worldwide is smoked in China. The Chinese consume an estimated 1.7 trillion cigarettes per year – or 3 million cigarettes every minute. In countries where health warnings on cigarette packages are mandatory, we tend to assume a universal knowledge of the risks of smoking. Yet, according to the World Health

Organization (WHO), one survey found that 60 per cent of Chinese adults did not know that smoking can cause lung cancer, while 96 per cent were unaware it can cause heart disease.

In the late 1980s, Korea opened its market to foreign tobacco firms. As advertising increased, cigarette consumption rose dramatically. In a single year after Korea lifted the ban against American tobacco, smoking among Korean teenagers rose from 18 to 30 per cent; among female teens it rose more than fivefold, from 1.6 to 8.7 per cent. The Republic of Korea is the eighth largest cigarette market in the world, with an annual volume of 100 billion cigarettes.

About 51 per cent of men smoke in Japan. This figure has dropped from the 1980s, but it is still very high for a developed nation. A survey in the early 1990s found that 44 per cent of male physicians smoke in Japan.[8]

By way of contrast with Pacific Rim countries, in Canada and the United States about 30 per cent of adult males and slightly fewer adult females smoke. In Great Britain, 28 per cent of men and 26 per cent of women smoke. By the mid-2020s, it is predicted that only about 15 per cent of the world's smokers will live in developed countries, as there will be a shift in the use of tobacco from developed (wealthy) to developing (poor) countries.

Along with cigarette smoking, high blood pressure is the other reason for the high rate of multi-infarct dementia in the Pacific Rim. High blood pressure is so prevalent in China that this one condition is responsible for up to 50 per cent of all deaths.[9] Interestingly, in China high blood pressure causes five times more strokes than it does heart attacks. The incidence of strokes in China is twice as high as that in Europe, America, and Japan.[10] Between 1 million and 1.5 million Chinese die from stroke each year.[11]

In the United States, although hypertension and related heart diseases account for 40 per cent of *all* deaths, there are about two and half times as many deaths from heart attack as from stroke.[12]

Pacific Rim countries suffered such a high number of strokes in the 1970s partly because hypertension was very often undiagnosed, especially in rural areas, and until the late 1970s, even if diagnosed, was rarely treated.

More recently, Alzheimer's disease has become a much more important cause of dementia in the Pacific Rim, rising to meet, and in many cases exceed, the rates of dementia from other causes. As we shall see, the rise in the AD rates coincides with a rise in the rate of meat-packing and meat consumption.

INCIDENCE AND PREVALENCE OF DEMENTIA

It is useful to be aware of the distinction between two terms used to describe how widespread a disease has become. *Incidence* is the number of new cases of a disease arising from a defined population per period of time. *Incidence may be expressed as new cases per population per year.* For example, the incidence of Creutzfeldt-Jakob disease (CJD) in the United States is thought to be one in a million. That means that in any given year, the number of new cases of CJD in the U.S. would be about 290 (or one new case for every million people in the U.S.).

Prevalence, on the other hand, is a snapshot in time. It represents the number of all cases – new and old – that exist in a population at a given time. *Prevalence rates are given as cases per one hundred of a given population.* Put another way, prevalence represents the disease burden on a population at any given time.

This means that for diseases that are rapidly fatal, such as CJD, the incidence and prevalence are about the same. As new cases arise in a period of time, a similar number of victims die within the same period. However, for diseases that are not immediately fatal, such as diabetes, the prevalence rises from year to year as the number of cases accumulate in a population. This example assumes that the incidence remains the same – that is, a similar number of new cases arises each year in the population.

In an important review of medical literature written between 1966 and 1999, G.H. Suh and A. Shaw came to some startling conclusions about the incidence and prevalence of dementia:

- First and most surprising was the change in the prevalence of dementia in the Pacific Rim (Korea, Japan, and China). In the 1980s, vascular dementia (VaD) was more common (MID is a specific type of VaD). By the 1990s, Alzheimer's (AD) had become more prevalent. A 1982 study in Japan revealed that for every hundred people, there were 2 cases of vascular dementia and only 1.2 cases of Alzheimer's disease. The ratio of the prevalence of VaD to AD was 1.7; that is, there were 1.7 cases of vascular dementia for every case of Alzheimer's disease. By 1999, the prevalence of AD in males had risen to 2.0 per 100 people to match that of VaD. The rate of AD in females had increased to 3.8 per 100 people with a VaD rate of 1.8. In other words, Alzheimer's disease is on the rise.

- Second, the prevalence of all dementias in Africa was relatively low. In 1997, the prevalence of both AD and VaD were less than 1 per 100 people (0.7 and 0.3 respectively).
- Third, studies from India were contradictory, but rates of AD were also low. No studies have found rates of AD greater than about 1 to 1.4 per 100 people.
- Fourth, American and European studies all showed a higher rate of AD than VaD. These studies all show rates of AD between 3.1 and 8.7 per 100 people, while those of VaD range from a low of 0.1 up to 3.0 in one study. The ratio of the two VaD/AD ranges from less than 0.1 to a high of 0.5.[13]

Why is Alzheimer's disease on the rise in the Pacific Rim? Why do India and Africa have low rates of dementia? Why are rates of AD so high in North America and Europe? In order to understand this data we will look carefully at Alzheimer's disease and its surprisingly short history, as well as looking at other diseases that appear to be related to it. Our unravelling of this medical mystery will take us into the rainforests of Papua New Guinea, the highlands of Scotland, and the meat-packing plants of the United Kingdom and North America.

CHAPTER 2

DEMENTIA: THEN AND NOW

No one is born demented, but sadly far too many people today die demented. Many people are surprised to learn that this was not always the case. How has the wise elder of yesteryear become today's confused denizen of a locked ward?

To understand what happens as a mind slips into dementia, it is worthwhile to look at what a normal brain does, and how it grows and matures.

THE BRAIN IN FORWARD GEAR: WATCHING BABY GROW

If you have an infant in the family, you can see the automatic responses at work. Place a finger in the newborn's mouth and she will automatically start to suck. Stroke her cheek and she will turn her head so that the stroking finger goes into her mouth. These two reflexes are called the sucking and rooting reflexes. These reflexes disappear shortly after birth as the intellect starts to develop, never to recur except in those unfortunate patients with end-stage dementia.

Most organs in the body are fully developed and function at birth. Certainly muscles, along with everything else, will get larger and stronger as a child grows, but they will not fundamentally change in what they do. The same is true of the heart, lungs, and most other organs. From birth onwards they continue to perform exactly the same tasks.

The brain, however, is different. Perhaps due to its incredible complexity, it requires far more than nine months in the womb to develop fully. When a child is born, the only fully functional parts of the brain are those responsible for basic life-sustaining functions: breathing, heart rate, digestion, sleeping, and waking.

As the growing brain matures, a child begins to acquire the mental skills that will enable him or her to function independently in the world. These skills come on-line, as it were, at different ages as different areas of the brain mature. At birth the amygdala, an almond-shaped area near the centre of the brain, is already developed, enabling the newborn to perceive emotions such as fear or well-being. But the area where consciousness resides has not yet matured, so the infant does not remember these emotions.

The next area to mature is the parietal area, located roughly behind the ears. This is the part of the brain that allows us to view our body as different from the rest of the world. We all know that we can move our arm simply by willing it to move, but we can't move that pencil sitting on the desk beside us using the same method (so-called telekinesis aside). A newborn doesn't know this. The newborn does not have an innate knowledge of where he ends and the outside world begins. This knowledge is acquired at three to four months of age. During this time the babe in diapers will often stare at his hands for long periods as he realizes, "Hey! I can control these things!"

At about six months of age the prefrontal lobes located just above the eyes start to function with the beginning of cognition and a sense of self. A child at this age begins to be aware of himself.

Also at six months, the limbic system – which includes the amygdala, hippocampus, thalamus, and hypothalamus, all relatively small structures located in the central lower areas of the brain – starts to mature and the first glimmerings of reason begin to appear. By one year, as this area matures, a child when offered a choice of two objects will select one rather than trying to grab both. Now you can play a new kind of game with baby. Place a toy under a blanket, and the child will realize that it has not disappeared, and will make an effort to retrieve it from where it is hidden.

By one year the reasoning part of the brain begins to take command of the limbic system, or the emotional part. As this is happening, babies love to play peekaboo – they are beginning to realize that Mother doesn't really disappear behind her hands. This early glimmering of understanding brings smiles both to the child and its mother. As this area matures, the child begins to lose interest in its reflection in a mirror, realizing that that's not really another child in there. Place a dab of jam on her cheek, and she will rub the cheek, not the mirror. This phase marks the beginnings of consciousness that is the awareness of one's self. This attribute – awareness of self – is so central and ingrained in the human brain that it appears to be the last intellectual trait that vanishes as one sinks into the abyss of dementia.

Between twelve and eighteen months the areas involved in speech mature. Wernicke's area, which allows one to understand speech, matures first. Now you can play the "Where's your nose?" game. The child understands and can point to her nose. She may not yet be able to say the word *nose*, but she's gathering words at a furious rate. Several months later, Broca's area, which allows one to produce speech, matures – and suddenly baby is talking. This asymmetrical development may cause frustration in the child who knows what she wants to say but can't say it. Appreciating this developmental quirk may help beleaguered parents understand the "terrible twos."

Can you remember the first time you looked at your mother the day you popped out of the womb? How about the excited congratulations you heard when you took your first steps? It would seem that these important life events should be forever etched in one's mind – but not a single person remembers them. The reason we can remember almost nothing before age three is that the hippocampus does not mature until we are about three years of age. The hippocampus, that piece of the brain located very close to the amygdala, is responsible for laying down and storing long-term memory.

Certain areas such as the frontal lobes do not fully mature until adulthood. These areas are responsible for our ability to focus on a specific task and, to a major extent, turn off all other sensory and mental stimuli

•

while we do the job at hand. This late maturation may help explain why more mature adults sometimes see teenagers as flighty and fickle. If your teenagers get hold of this information, they'll tell you that this is why it's hard for them to settle down and do their homework, or why they're really not the best person to paint the fence.[1]

THE BRAIN IN REVERSE GEAR

When the brain begins to fail as happens in dementia, the same sequence of events occurs – but in reverse. Victims lose the ability to focus on a task, a problem that initially may be reflected in their poor performance at work. An accountant finds that jobs that once were routine become difficult, if not impossible. A widower who has lived successfully on his own for years now finds that planning and preparing a simple meal confounds him.

A retired academic can no longer read a novel as her attention wanders off the page. Later she begins to get angry and frustrated as she loses the ability first to produce coherent speech, and later to comprehend what is being said to her. Just as the young child gathers words at a furious rate, with the sequence unwinding, the Alzheimer's victim loses words with frightening momentum.

As this process marches on relentlessly, the victims' ability to distinguish between themselves and the outside world seeps away. Finally, family and friends witness the sad spectre of a once-vigorous, dynamic individual now in diapers, unable to walk, speak, or feed himself, lying in a bed with side rails, staring mutely at his hand.

THE LONG AND THE SHORT OF IT

A middle-aged woman remembers her mother's decline into dementia. "At first I couldn't understand why she was so angry with me. I thought I must have been doing something wrong. I realize now how frustrating it must have been for her to be losing control of her life – and her mind." A charming gentleman at a party remembers all the details of his fascinating life working as a ventriloquist – but tomorrow will not remember having met the person to whom he told the story.

Those with first-hand experience of Alzheimer's often lament that the true tragedy of the disease is that it robs people of their memories. Is anything sadder than a parent who cannot recognize a child, or a spouse who does not know his loving soulmate of fifty years?

Memory is basic to survival. Primitive forms of memory can be found in even the simplest, single-celled animals. Any living entity requires some form of memory in order to avoid the very hot areas, for example, or to move from where food is scarce to where food is plentiful. All learning is dependent on memory, from knowing not to put your hand on a hot stove to knowing how to stand up when you get out of bed in the morning.

We can divide the function into two broad categories – short-term, or working, memory and long-term memory.[2]

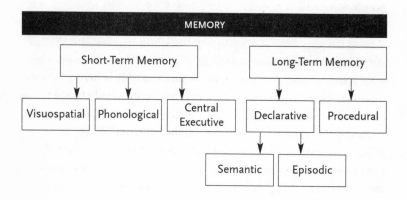

Short-Term Memory: Do You Want Fries with That?

We use short-term memory when looking up the number of our local pizza delivery place. Short-term memory has both a limited capacity and limited duration. We can hold about seven numbers – incidentally the length of a telephone number – in short-term memory. But short-term memory is easily lost. If, on your way to the phone, you happen to notice a newspaper headline, by the time you have read it, the phone number is forgotten.

Working memory can be further broken down into three parts. Psychologist Dr. Alan Baddeley, now at the University of Bristol, originally proposed the idea of the working memory as a system comprising multiple components in the mid-1970s. These components are:

- the phonological rehearsal loop;
- the visuospatial sketchpad; and
- the central executive system.

The first two components are relatively passive systems responsible for temporary storage of verbal and visual information respectively.

The *phonological loop* is the "voice in your head." It is how we learn language by repeating in our heads, or how we keep going over the pizza place phone number as we reach for the phone. When you meet new people at a gathering, you may find yourself repeating names in your head to help you remember who's who – that's your phonological loop at work. If you don't start saying the name in your head immediately after hearing it, you are likely to forget it within seconds.

The *visuospatial sketchpad* is our "mind's eye." It is here where we imagine and manipulate visual images. How will this red-striped tie in the men's clothing store look with the grey suit that I have at home? What size of carpet will fit that space in the bedroom? When the waiter asks you, "Do you want fries with that?" your visuospatial sketchpad lets you picture your meal with and without fries so you can decide whether that tempting taste is worth the extra calories and fat.

The *central executive system* controls what is happening in the other two parts – the phonological loop and the visuospatial sketchpad – of the working memory. This centre orchestrates resources like memory, language, and attention to achieve an objective – either long-term or short-term. This central system is very active, selecting, initiating, and terminating processing routines (e.g., encoding, storing, and retrieving). It's the central executive system that allows you to call on your phonological loop when you're talking with your wife, so that you can recall what she has just said and rehearse what you are about to say. If she asks you, "Do you think I should cut my hair?" your central executive system ensures that your visuospatial sketchpad takes over. (Then it's just a matter of good judgment whether you answer truthfully.)

We use the central executive system in performing tasks that require planning or sequencing – we turn the doorknob, open the door, and walk through the doorway. People with learning disabilities and conditions such as attention deficit hyperactivity disorder, along with those who are simply disorganized and inefficient, may be said to have "executive dysfunction." However, this central processing system is severely impaired in people with AD. Even the simplest of tasks require sequencing. People with AD often have to be fed, not because they cannot perform acts such as putting food on their forks or conveying the food to their mouths. They require help because they cannot put these two simple acts together in the right order.

Long-Term Memory: Where Were You September 11?

The other major component of memory is long-term memory, which is divided into two broad categories that answer the questions "What is?" and "How to?"

- The "what is" memory is known as declarative memory, and consists of a huge collection of facts learned, things seen, language acquired.
- The "how to" memory is known as procedural memory.

You walk down the street and see a person pedalling vigorously on a device consisting of metal tubes and two large wheels. You don't have to say to yourself, "What is that?" Your declarative memory lets you know instantly that the person is riding a bicycle.

Declarative memory is further divided into semantic and episodic memory. *Semantic memory* consists of the enormous collection of words, facts, objects, concepts, people, and places that comprise "what we know." It is what is involved in learning language and abstract concepts such as mathematics, and how we apply the knowledge we gain.

Strokes that affect the part of the brain known as the temporal lobes may cause a fairly rare form of dementia called semantic dementia.

People with this disorder provide us with insights into how semantic memory is organized. Someone with semantic dementia may be unable to recall specific words. This inability to name an object is called anomia. For example, if a person with anomia is shown a screwdriver and asked to name it, he will not be able to do so. If asked to guess, he might say, "It's a hammer." However, he will never say, "It's a bread stick," even though a screwdriver looks more like a bread stick than like a hammer. It appears that we store memories in categories. So although the word for screwdriver is gone, the individual still knows that the word is stored in the tool category.

Some people who have suffered from strokes in another area of the parietal lobe of the brain can lose the ability to do even the simplest of mathematics – a condition called dyscalculia. Patients suffering from dyscalculia understand numbers as nouns but not as abstract ideas. If asked, "How old are you?" a victim will tell you her correct age. But if you then ask, "How old will you be one year from now?" she will be unable to answer.

The capacity of semantic memory is vast. In a classic 1973 experiment called the Standing Experiments, Canadian psychologist Dr. Lionel Standing showed his subjects a series of several dozen slides of either pictures or words, allowing only five seconds for each. Three days later, he showed the subjects pairs of slides, one from the original group, the other with a picture or word that the subject had not seen before. Subjects were asked to identify which one they had seen before in each pair. The subjects almost always got them all right, so the experimenter began to increase the number of pictures. He found that even when he used ten thousand pictures, subjects still got almost all of them correct. Standing concluded that, for all practical purposes, "there is no upper bound to memory capacity."[3]

Episodic memory, the other component of declarative memory, allows people to recall and replay personally experienced events from their past. This situation has been likened to taking a photograph, in that experiences happen in a second, and during that time we can store not only the event, but also all sorts of ancillary information surrounding the event. Furthermore, we can reinforce episodic memory through conscious mental rehearsal, review, and reflection – so looking at photos of your summer vacation helps you to remember even details that do not appear in the photograph.

Emotion also plays a huge role in laying down these memories. Ask almost anyone living in the United States what he or she was doing on September 11, 2001. Not only will they be able to recall where they were, but they will also remember details such as what they were doing, what they were wearing, who else was there, and a raft of other minutiae. The emotional impact of the tragic and frightening attacks on the World Trade Center and the Pentagon cemented in memory many personal details of the day for those who were deeply affected.

It can be said that semantic memory is what we know, while episodic memory is who we are. Young children attending the same kindergarten may have similar semantic memories but very different episodic memories. All of the children may learn what a sandbox or a dump truck is and remember that information forever, but one may remember that she got sand in her eyes, while another remembers losing the battle for control of the yellow dump truck. Yet another may remember that the teacher was named Ms. Jones, and she gave him a red drink.

Procedural memory is our "how to" memory. This function allows us to recall how to perform physical tasks – everything from walking, to tying our shoelaces, to playing the trumpet. Once established, this memory is remarkably durable. The old adage makes perfect sense – it truly is just like riding a bicycle. You may not have done it for twenty years, but all the skills come back within seconds. Some things that we assume are stored in declarative memory are really in procedural memory. You can try an interesting experiment – turn the phone 180 degrees so that the zero is on top of the dial pad. Now dial a number that you are very familiar with. You will be able to do it, but it will feel awkward and take much longer than when the phone is in its normal position. Procedural memory is retained longer than any other form of memory in AD.

Amnesiacs are unable to add to their declarative memory stores – either episodic or semantic. Amnesiacs have normal memory for events that happened before their trauma, but they cannot lay down any new long-term memories. However, they are able to learn new rules or procedures, clear evidence of the separate functions of different types of memory.

Learning is the process of moving information from short-term memory into long-term memory, and this process is severely affected in AD. A person with AD may recall in great detail her own wedding that happened fifty years ago, but will be unable to recall her grandchild's wedding that took place yesterday.

DIAGNOSING ALZHEIMER'S DISEASE

Mild cognitive impairment, usually the first sign of Alzheimer's disease, is a term that has been used only since the late 1990s. People with mild cognitive impairment are unable to form memories for events that just happened. Mild cognitive impairment may be caused by depression or other disorders. But when no such cause is found, at least 80 per cent of cases develop into full-blown Alzheimer's disease.

Another early symptom of AD is slight shrinking of the hippocampus, where we lay down long-term memories. Scientists are now working on a computer chip that may one day replace living neurons in the hippocampus with silicon ones – a procedure that may aid in assisting not only AD victims but also those suffering from language and memory problems that result from stroke or epilepsy. But until that day comes, nothing can stop the steady mental decline associated with AD.

Although the definitive diagnosis of Alzheimer's still relies on actually seeing distinctive lesions in the brain under the microscope (during autopsy), more than 99 per cent of AD diagnoses are based on first assessing if a dementia is present. If it is, then the next step in diagnosis consists of ruling out the non-AD causes of dementia. Once this has been done, the diagnosis can be made with a fair amount of certainty.

To understand how this process works, consider a typical patient taken to the doctor by a son concerned about his mother's deteriorating memory. The initial interview is held in the doctor's office with the parent and son both present. Questions are always directed at the patient. The son's role is to confirm the accuracy of the answers and help if the patient cannot answer.

The doctor first tries to put the current problem into a temporal context. This means he or she will want to know if this memory loss came on quickly or slowly. Was there a single event such as an operation or a major medical condition such as a heart attack or stroke after which her memory got much worse or has this memory loss been gradual?

Is there a family history of similar conditions and, if so, who was affected and at what age did they begin to show symptoms? This is important, as there is a form of AD that is definitely inherited. Has the patient ever had a problem with drugs or alcohol? These can predispose one to memory problems. The use of prescription medications, especially drugs such as sedatives, tranquilizers, and sleeping pills, can seriously affect mental function, especially in the elderly, so the doctor will note any use of these. Does the patient have a psychiatric history? Conditions such as schizophrenia, manic-depressive disease, and depression can all mimic AD.

After taking a history, the doctor does a thorough physical examination, with special emphasis on the sensory organs. Older patients whose vision or hearing is poor may appear confused or demented when their problem is that they can't see or hear properly. In these cases, new glasses or a hearing aid can often cure their "dementia." The sense of smell is almost always decreased in AD; however, loss of this sense is so common in the elderly from a variety of other causes that this finding is of little value in diagnosing AD.

The next step is to evaluate mental function. The Folstein Mini-Mental State Examination (MMSE) is the most widely used screening test of cognition in older adults. Dr. D. William Molloy, an internationally recognized Alzheimer's researcher, is the director of the Memory Clinic and the Geriatric Research Group in Hamilton, Ontario. He has written administration and scoring guidelines called the Standardized Mini-Mental State Examination, usually referred to as the SMMSE.[4]

The SMMSE can be used to determine what stage of the disease the patient currently is at, to differentiate between the kinds of dementias, and to assess whether a patient has responded to treatment. This test is reproduced here, along with the scorecard.

The doctor says to the patient, "I am going to ask you some questions and give you some problems to solve. Please try to answer as best as you can."

1. (Allow 10 seconds for each reply)   Maximum score
   a) What year is this?   1
   b) What season is this?   1
   c) What month of the year is this?   1
   d) What is today's date?   1
   (Answers can be off by one day)
   e) What day of the week is it?   1
2. (Allow 10 seconds for each reply)
   a) What country are we in?   1
   b) What state/province/county are we in?   1
   c) What city/town are we in?   1
   d) What building are we now in?   1
   e) What floor are we on?   1
3. "I am going to name three objects. After I have
   said all three names, I want you to repeat them.
   Remember what they are because I am going to
   ask you to name them again in a few minutes."
      The words are said slowly at one-second intervals.
   "Ball. Car. Man. Please repeat the three items
   for me." (Score 1 point for each correct answer)   3
4. "Spell the word WORLD." (You may help the
   subject with the spelling.) "Now spell it backward
   please."   5
5. "Now what were the three objects that I asked you
   to remember?" (Score one point for each correct
   answer regardless of order, allow 10 seconds)   3
6. Show subject your wristwatch. Ask, "What is this
   called?" Accept wristwatch or watch. Clock or time
   are not acceptable.   1
7. Show pencil. Ask, "What is this called?" Accept
   only pencil.   1
8. "Please repeat this phrase after me: No ifs, ands,
   or buts." Answer must be completely accurate.   1
9. Say, "Read the words on this piece of paper and then
   do what it says." Hand subject a piece of paper with
   CLOSE YOUR EYES on it.

Only correct answer is if the subject closes their
eyes. It does not matter if they read it aloud or not.   1

10. Ask if subject is right or left-handed. If right-handed
take a piece of paper, hold it up in front of subject
and say, "Take this paper in your right hand, fold the
paper in half with both hands and then put the
paper down on the floor."

Takes paper in correct hand   1

Folds paper in half   1

Puts paper on floor   1

11. Hand subject a pencil and a piece of paper and say,
"Write any complete sentence on that piece of paper."
Score 1 point if they write any sentence that makes
sense. Ignore spelling.   1

12. Place a piece of paper on the table. Make a simple
design such as two interlocking pentagons on the
paper. Give the subject a piece of paper and a pencil
and say, "Please copy this design." Score one point if
the design is essentially correct.   1

                                                    _____

Total Test Score   30

On this test most elderly people should score 30 out of 30. Scores
between 26 and 30 are considered normal. Scores between 18 and 26
indicate mild cognitive impairment. Below 18 indicates a significant
deficit in reasoning ability.

After doing this assessment, the doctor will probably order a battery of
lab tests to rule out other potentially treatable causes of dementia, such
as low or high thyroid levels, anemia, vitamin B12 deficiency, kidney failure, liver disease, AIDS or syphilis, or calcium or electrolyte imbalance.
Various X-ray or other imaging studies such as CT scans and MRIs may
be ordered, but their value in making a diagnosis is not very great.

If the subject scored poorly on the SMMSE and all the laboratory tests
are negative, the diagnosis of AD or a similar dementia is made.

Currently people with signs or symptoms suggestive of AD are placed
in one of three categories:

- Definite Alzheimer's disease: This diagnosis can be made only after death, and is reserved for those people who show characteristic brain lesions at autopsy and had behaviours indicative of AD when alive.
- Probable Alzheimer's disease: This is the most common diagnosis. This would be a person who has shown progressive decline in mental function with no other brain disease or other condition to account for the changes.
- Possible Alzheimer's disease: This diagnosis is for those people in the early stages of the disease where their memory or behaviour has not yet reached the stage where a diagnosis of probable AD can be made.

This system, if used by persons trained in the diagnosis of the disease, is remarkably accurate. More than 90 per cent of those diagnosed with probable AD reveal characteristic changes in their brains if they undergo an autopsy after death.[5]

"AN UNUSUAL DISEASE OF THE CEREBRAL CORTEX"

1906. Tübingen, Germany. Dr. Alois Alzheimer, a German physician, rises to speak at the 37th Conference of South-West German Psychiatrists.

He tells the assembled crowd about a patient of his, a woman in her early fifties named Auguste D., whom he had observed since 1901. As one of her first disease symptoms, she had shown a strong feeling of jealously toward her husband. "Very soon she showed rapidly increasing memory impairments; she was disoriented carrying objects to and fro in her flat and hid them. Sometimes she felt that someone wanted to kill her and began to scream loudly. . . . After 4 years of sickness she died."[6]

Alzheimer then begins to describe the terribly damaged brain he had seen after her death.

Alois Alzheimer (1864–1915) had embarked on his medical career in December 1888 as assistant physician at the Municipal Hospital for Lunatics and Epileptics in Frankfurt, Germany, where he was later promoted to second physician (senior physician). There he began research on the cortex of the human brain. Wishing to combine research with clinical practice, he became research assistant to Emil Kraepelin at the Munich medical school, and created there a new laboratory for brain research.

A leading neurologist, Alzheimer published works on epilepsy, brain tumours, syphilis, and hardening of the arteries. He developed a particular interest in studying the brains of patients who had died, and became skilled at correlating the clinical course of his patients with the pathology he observed in their brains.

But in November 1906, he reveals for the first time a completely new finding, an "unusual disease of the cerebral cortex." Mrs. Auguste D., he tells the psychiatrists, had experienced progressively worse symptoms over the several years he had observed her. Her difficulties included severe memory problems, progressive cognitive impairment, speech and perception problems, hallucinations, delusions, and social incompetence. He recalls that when asked her name, she could reply, "Auguste," but when asked her husband's name, she replied again, "Auguste." As time went on, she could write the alphabet only up to G, and then later not at all. She had died earlier that year in an advanced state of dementia: incontinent, incoherent, and immobile. Upon her death, Alzheimer had examined her brain, using a silver staining technique and new distortion-free microscopes that allowed him to view the nerve cells.

To the attentive conference assembly, Alzheimer describes what he had seen: the shrunken brain, and outside and around the nerve cells unusual dense deposits of cellular material. Inside the nerve cells, he tells his colleagues, he had found threadlike, spindle-shaped twisted bands of fibres.

Today we know these dense deposits as neuritic plaques, and the twisted fibres as neurofibrillary tangles. The observation, at autopsy, of these plaques and tangles is still required today to confirm a definitive diagnosis of Alzheimer's disease.

In 1910, Alzheimer's mentor and partner, Dr. Emil Kraepelin, suggested that this new disease be named for its discoverer. Meanwhile, other doctors began to report similar cases. Kraepelin may have rushed to name the disease because he wanted the prestige for his Munich laboratory. Some claim that Kraepelin was seeking support in his dispute with Sigmund Freud over whether there was a biological component to mental illnesses.

(While Freud favoured the idea of "constitutional predispositions" to mental illness, Kraepelin supported a "hereditary taint.") Nevertheless, Kraepelin and Alzheimer had worked closely, and Kraepelin felt it appropriate to honour Alzheimer's clinical and scientific work on presenile cases. Ever since then, the name *Alzheimer* and dementia have been synonymous.[7]

Cases of dementia associated with aging were not unheard of before Alzheimer's time; however, they were exceedingly rare.

Kraepelin was one of the most influential psychiatrists of his time. He created the first modern classification of mental diseases, and his distinctions between schizophrenia and manic-depressive psychosis remain valid today. Among Kraepelin's outstanding students was Alfons Maria Jakob (1884–1931), who became famous for describing Creutzfeldt-Jakob disease (CJD).

In 1904, Kraepelin published a collection of his *Lectures on Clinical Psychiatry*,[8] including a chapter on senile imbecility. Of the several cases mentioned, at least one describes a patient, a seventy-two-year-old woman, who may had an Alzheimer-type dementia.

His description notes: "She is quite unconscious of the grossest contradictions in her statements of time. Thus she asserts that her daughter is two years younger than she, that her father is sixty, when she has just given her own age as sixty, that her child is three years old, and so on."[9] He speculates that there may exist a form of senile imbecility that may be distinct from the imbecility that one sees in the elderly as a result of long-standing mental disease.

Within a short time after Alzheimer's lecture, other physicians began to note similar brain damage at the time of autopsy. Tracking the number of cases of Alzheimer's disease that appeared in the medical literature in the years following Alzheimer's lecture reveals a significant pattern. From a single case mentioned in 1908, we find a slow progression upwards, with a rapid increase in the late 1940s. We see another leap in recorded cases to 648 in 1978.

These listings do not represent all cases of Alzheimer's disease, only the number of cases cited in studies in the medical literature in any given year. For example, in all the studies published in 1953, the total number

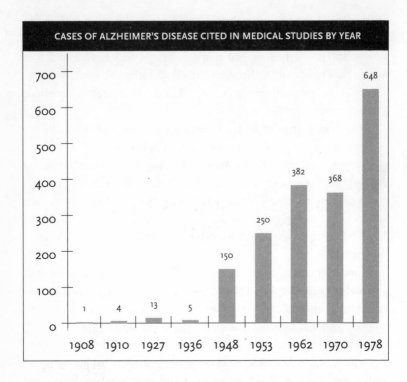

CASES OF ALZHEIMER'S DISEASE CITED IN MEDICAL STUDIES BY YEAR

of AD patients mentioned was 250. The figures show that physicians interested in researching the disease were finding ever-larger numbers of cases to investigate in the years following Alzheimer's original paper.[10]

The numbers today are so huge they could not be plotted on this chart. A single study might involve several institutions in several cities and the number of patients in the thousands.

Alzheimer's disease has become more and more common – to the point that today the dementias of aging are considered the norm rather than the exception. But, as we will see, it was – even in fairly recent times – so rare that the greatest medical observers of all time did not even mention it.

DOCTORS AND MEDICINE
IN THE OLD DAYS

*"Learn to see, learn to hear, learn to feel, learn to smell and
know that by practice alone can you become experts."*
– Sir William Osler, physician and writer (1849–1919)

A grand old matriarch of our acquaintance, noted for keeping a tight grip on her money, was once – in the days before medicare or health insurance – so sick that she *almost* called the doctor. Luckily she recovered without having to part with her hard-earned cash, and ended up years later leaving her estate to her heirs.

Notwithstanding our thrifty friend, for thousands of years people have been paying to consult physicians. Despite the old saw about prostitution, we can probably safely say that medicine is the oldest profession. Yet until very recent times, doctors enjoyed only a modest success rate in curing their patients.

Today people are willing to spend large sums to see a physician or to support health care with massive amounts of tax dollars so a physician can assist them to get better. Physicians earn their living today by their ability to heal. This may require medication, psychotherapy, or in severe cases hospitalization and intravenous drugs, electronic monitoring, sophisticated laboratory tests, or even surgery.

Almost all of these treatments have been developed in the last one hundred years. In the nineteenth century, there were no antibiotics, so what would be a minor infection today was often a fatal affliction back then. Psychiatry was in its infancy, so mental illness was either ignored or, worse, treated by locking the unfortunate victim in places so awful they were considered appalling to liberal thinkers at the time. Society hadn't yet figured out any other way to protect itself from the perceived threat that madness posed.

Surgery was done without the benefit of anaesthesia, so the mark of a great surgeon was the speed with which he could perform an operation. Some military surgeons of the Napoleonic era were able to amputate a limb in less than five minutes, a procedure that takes about three hours today. It was not only the fear of terrible pain that kept people away from surgeons. Antisepsis was unknown. If a patient survived the pain and brutality of the operation, the odds were greater than 90 per cent that he or she would die anyway as a result of an infection.

Despite knowing this, people still paid large sums to consult a physician. Why? Prior to about 1900, why were people willing to spend money to see a doctor? The answer lies in what physicians actually did, and how they were trained two hundred years ago.

Before about 1925, there were no standards for what constituted a reputable medical school. In the nineteenth century, admission to a faculty of medicine was confined to gentlemen (women – gentle or otherwise – were not admitted to medical schools in significant numbers until the 1950s) who could demonstrate ability in the classics. An entrance exam from 1898 included the requirement that the applicant translate a passage from Latin to English and another from English to Latin, and to be familiar with the works of Virgil, Shakespeare, and Poe.

The applicant also had to be conversant with classical history (Roman and Greek) as well as European and North American history. He had to know the rules of grammar, syntax, composition, and rhetoric in both English and Latin. In the sciences he was expected to solve problems in algebra, physics, geometry, and arithmetic. The only life science required was botany. A question in botany from that time reads: "Show the differences between the pod of a pea and that of a radish. Also show the varieties of shape in the fruit of the mustard family."[1] (Try asking that of your family doctor today.)

The curriculum varied widely among different medical schools, but certain things were considered basic to all reputable institutions. In the late 1800s, the ability to read and write was considered necessary for admission to some schools. Since Hippocrates, medical lore had been an oral tradition passed along from mentor to student, but the rapidly increasing volume of medical knowledge in the nineteenth century finally changed the nature of a doctor's education. The length of training varied from about three months to more than four years.

MEDICAL TRAINING IN 1900

What was medical school like in 1900? What was expected of the student? Then, as now, he would need to gain knowledge, first by attending formal lectures and seminars, and secondly by acquiring practical knowledge at the bedside. In fact, our modern word *clinic* is derived from the French *clinique*, originally "bedridden person."

But then, as now, the student could not master the sheer volume of material merely by these methods – so he would need to learn also by reading. Until very recently every physician owned a library of classical medical texts, which he or she would consult on a regular basis. This began to change in the mid-twentieth century with the advent of medical journals that allowed a doctor to remain current without having to purchase the newest edition of his or her relevant texts. It changed again in the late twentieth century when the Internet provided even more up-to-date information.

In 1900, just as now in many countries, medical school took four years. The curriculum, however, was very different. A student entering medical school in 1900 would spend about one-third of his time studying anatomy. This served two purposes. First, he would learn how the human body is constructed, all of its myriad parts, where they were, and how they related to each other. Second, he would learn a new language. The study of anatomy taught him most of the nouns of this new language. Every bone, vein, and nerve has a name; in fact, every bump on a bone has a unique name. This new language is one of the things that define a profession, the ability to talk to your peers in a language that only they understand, sort of a verbal secret handshake. The average English-speaking person has a working vocabulary of about five thousand to six thousand words. In the first two years of medical school a student will learn about ten thousand to fifteen thousand new words.

Medical students in the early years of training also studied physiology, or how the various systems in the body work. Students would learn about the way blood circulates, and how the respiratory and digestive systems work. By the end of the second year of study, a student would have a good knowledge of what the healthy human body was made up of and how it worked.

In order to pass exams, students needed to know their subject in an amazing degree of detail. A question from an anatomy exam at the time might read, "If a long pin were stuck into a person, entering at the bridge of the nose and exiting between the fifth and sixth cervical vertebrae, name at least one hundred structures that the pin would pass through."

(The study of anatomy continues to provide tools for the torture of medical students. When I was in medical school, the professor of anatomy used to torment students by carrying all seven human carpal bones in the pocket of his lab coat. These are the tiny bones, each smaller than a sugar cube, that make up the wrist. The professor would reach into his pocket, pull out one bone, toss it up in the air, and demand that the terrified student identify it before it came back down.)

The next phase of the medical student's education encompassed the topic of what can go wrong – the study of pathology. In the days before sophisticated laboratory tests, ECGs, MRIs, CT scans, and all the modern aids to diagnosis, a student had to become proficient at diagnosing diseases almost solely by observation. He would learn to use all of his senses – much as the great physician Sir William Osler urged students to do. A student needed to recognize by:

- sight: that certain forms of anaemia caused the fingernails to take on a greenish colour;
- scent: that a patient with advanced diabetes smelled of fresh apples;

- taste: that when he (dauntlessly) licked the skin of a baby with cystic fibrosis, it tasted salty;
- feel: that a touch to the arm would reveal the doughy feel of the skin associated with dehydration;
- hearing: that the lower lungs of the patient who was in congestive heart failure produced fine cellophanelike crackling noises.

In short, by the time a student had completed his training, he had become a magnificent observer. It is no accident that it was a medical doctor trained in the late 1800s, Sir Arthur Conan Doyle, who created Sherlock Holmes, the fictional character who personified the deductive power of acute observation. The character of Holmes was closely modelled on Dr. Joseph Bell, a well-known physician of the time who was one of Doyle's clinical instructors at Edinburgh University. Noted for his genius as a diagnostician, Bell instructed his students that a doctor should use all his senses to find the cause of illness. He advocated that doctors not just look at a patient, "but feel him, probe him, listen to him, smell him."

In his autobiography, Conan Doyle relates this anecdote about his instructor Dr. Bell:

> In one of his best cases he said to a civilian patient: "Well, my man, you've served in the army." "Aye, Sir." "Not long discharged?" "No, Sir." "A Highland regiment?" "Aye, Sir." "A non-com officer?" "Aye, Sir." "Stationed at Barbados?" "Aye, Sir." "You see, gentlemen," he would explain, "the man was a respectful man but did not remove his hat. They do not in the army, but he would have learned civilian ways had he been long discharged. He has an air of authority and he is obviously Scottish. As to Barbados, his complaint is elephantiasis, which is West Indian and not British." To his audience of Watsons it all seemed very miraculous until it was explained, and then it became simple enough.[2]

In later years, Bell commented on the Sherlock Holmes stories written by his former student: "He shows how easy it is, if only you can

observe, to find out a great deal as to the works and ways of your inno-
cent and unconscious friends, and, by an extension of the same method,
to baffle the criminal and lay bare the manner of his crime."[3] Yes indeed,
if only you can observe.

Perhaps the most important subject a student would learn during his
pathology training concerned the natural history of disease. He learned
the progression of a disease from its onset until the patient recovers or
dies. Before the advent of effective therapies for most diseases, this was a
crucial part of the doctor's knowledge. The motto of a nineteenth-
century physician might have been "We probably can't treat it, but we
certainly can describe it."

People were willing to spend money to see a doctor partly because the
doctor could put a name to what was bothering them, and then tell
them what they could expect to happen next. Knowing the natural history of a disease, the physician could tell sufferers if their disease would
progress to death or disability, or whether they would get better. This
knowledge allowed patients and their families to plan their future.

Medical students then, as now, relied most upon this aspect of their
education – the observations of those who had preceded them. In a mere
four years of training, one could not be expected to see all of the thousands of diseases that afflict human beings. So in order to acquaint
themselves with conditions they had yet to see, and to confirm a diagnosis,
a student or practising physician would consult the appropriate text.

Early Medical and Psychiatric Observations on Dementia

Often called the greatest physician of the late nineteenth and early twentieth century, Sir William Osler (1849–1919) literally wrote the textbook
on medicine. Born in Canada and trained at the University of Toronto
and McGill University, he later became the first professor of medicine at
Johns Hopkins University. In 1905, he moved to England where he took
up the Regius Chair of Medicine at Oxford. Not only was he the best-
known physician in the English-speaking world, he was universally
acknowledged as the greatest physician of his time, and the most
influential physician in history. A brilliant observer and clinician, his

name is appended to many diseases and diagnostic signs such as Osler's disease, Osler's nodes, and Osler's sign.

Osler also wrote beautifully – and extensively. At one time, he noted that a bibliography of his writings included some 730 articles.

Osler is best known for publishing the first comprehensive textbook of medicine in 1892, an extraordinary work of medical scholarship. The opus *The Principles and Practice of Medicine* is regarded as the template for all modern medical texts. Twenty-one of the world's foremost physicians helped to expand the textbook, and in 1907 it was published under the title *Modern Medicine: Its Theory and Practice*, and was commonly called *Osler's Modern Medicine*. This text was a monumental undertaking. When completed, it ran to more than seven thousand pages in seven volumes, and represented the sum total of medical knowledge at the time.

The seventh and last volume, almost one thousand pages long, deals only with disorders of the brain and nervous system. This is where we would expect to find Alzheimer's disease – but dementia is mentioned only in the context of the dementia associated with late-stage syphilis and with certain forms of schizophrenia. No mysterious inexplicable mental deterioration is described.

Did Osler then gloss over the subject of dementia? Not at all. This is his observation of the cognitive impairment that occurs in late-stage syphilis:

> With this defective judgment *defective memory* is associated at an early date, so that the patient forgets the facts that everyone is expected to know – the day of the week, the month, the year, the name of the President or Governor, even the names of his family. The paretic (i.e., the patient with late-stage syphilis) at an early stage may be unable, offhand, to give the names of his children, or of his partner; his memory for recent events is often more impaired than his memory for things that have happened long since.
>
> It is curious to see how easily the physician may get at the evidence of an impaired memory. As a rule, it is best to begin by questioning the patient as to his age. If he states he is fifty-two years of age, and you ask him the year of his birth, he is

unable to give it; ask him how long he is married, he fails to state it correctly; if the wife in his presence supplies the answer, he is unable to state the year of his marriage. A simple sum of subtraction is quite beyond his mental powers. If asked to write his name and address, he will be able to do it correctly in the majority of instances, unless he is already in a very advanced stage. Ask him to write the names of some cities which have not come within his daily correspondence, and you will soon detect his inability to do so.

The things which he has known most thoroughly he may be able to do; those which require a little reflection he fails in. In the early stage the patient may write New York, Boston, and Buffalo correctly, he is apt to fail on Philadelphia, Rochester, and Constantinople, which is the stumbling block for many.

In every instance, of course, it is well to assure one's self that the patient spelled correctly before the disease came on. His memory may also be tested by putting before him the simplest sums in arithmetic. The ordinary additions he may be able to do, but ask him to multiply 8 by 13 or 7 by 8, or ask him to divide 120 by 15, and one quickly discovers the limitation of his knowledge. Far from being annoyed at his mistakes, the average paretic shows either an indifference to the test or considers it a huge joke.[4]

It's obvious that Osler and his colleagues thoroughly examined the mental status of patients. It is also remarkable in that if you showed this passage to any physician today they would swear that it is an excellent description of Alzheimer's disease. And yet the description refers only to those in late-stage syphilis. Nothing appears even vaguely resembling the other indicators of Alzheimer's disease. If the disease were present, is it likely that such a keen observer and recorder as the renowned Dr. Osler would have missed it?

Osler did mention insanity in an almost jocular fashion when discussing neurasthenia, a vaguely defined medical condition characterized by symptoms such as sensations of pain or numbness in parts of the body, chronic fatigue, headaches, fainting, irritation, and anxiety. He is quoted as saying, "In neurasthenia or insanity, '*cherchez la femme*' – woman is at the bottom of most troubles."[5] Perhaps today, to avoid offending both women and those suffering from vaguely defined symptoms, he may have been more careful with his words. Nevertheless, it's clear that dementia was not a common condition of the elderly as it is today.

Even in the 1935 edition of Osler's text, there is only one sentence dealing with the subject. "As the disease [cerebral arteriosclerosis] progresses, the mental state may fail, but, in contradistinction to the presenile and senile types of dementia, many of these people keep a clear mind, and there are none of the features of Binswanger's dementia, presenilins or of Alzheimer's disease."[6] That sentence is the only mention of dementia, with, of course, the exception of syphilis. Alzheimer's disease had been named less than thirty years earlier, and was still so exceedingly rare as to merit no further mention.

In a recent article in the *Canadian Medical Association Journal*, Dr. David Hogan compared the 1892 edition of Osler's *The Principles and Practice of Medicine* to the 1987 edition of Harrison's *Principles of Internal Medicine*. He noted how the two texts dealt with diabetes, ischemic heart disease, pneumonia, and typhoid fever. He found that except for the therapeutics sections (Osler felt there was no effective medicine at the time for any of these conditions), there are very few differences between their descriptions of these diseases.[7] In other words, the descriptions of all diseases that were in Osler's text are equivalent to those descriptions in the best modern texts. The striking exception is Alzheimer's disease, which is extensively described in Harrison's but is not mentioned at all in Osler's.

SEARCHING FOR DEMENTIA INSIDE THE BRAIN

Dr. William Boyd (1885–1979) first worked as assistant physician in the Derby Borough Asylum in Scotland. His sister's psychiatric illness may have influenced this choice of position. However, his major responsibility at the asylum was performing autopsies, which led to his future fame as a pathologist.

Pathology is the scientific study of disease in humans and other living organisms. Today, pathologists have access to blood and tissue samples of living people, and a wide range of analytical techniques (as anyone who watches *Da Vinci's Inquest* or other forensic television dramas knows). But in Boyd's day, pathologists depended largely on post-mortem dissection of bodies.

Perhaps the only physician to rival Osler in his impact on medical students, Boyd wrote the other landmark text of this era, *A Textbook of Pathology: An Introduction to Medicine*, widely recognized as the most comprehensive pathology text of the time. Well known for his colourful and lively prose (e.g., "Of all the ailments which may blow out life's little candle, heart disease is the chief."), Boyd ensured that his readers would not fall asleep while studying his text.

Despite having spent his early career in an asylum (which implied mentally disturbed patients), Boyd seems to have had no acquaintance with Alzheimer's disease. In Boyd's third edition[8] published in 1938, he makes no reference to AD or other neurodegenerative diseases of the brain.

Today, every pathologist recognizes plaques in the brain as a postmortem confirmation of a diagnosis of AD. But Boyd offers no histopathological (study under the microscope) evidence that plaques – a defining feature of AD – had been seen in brains at autopsies.

After reading both Osler's and Boyd's books, one is struck by their very thorough nature. If Alzheimer's or the other neurodegenerative diseases were endemic, it is difficult to imagine that both Osler and Boyd would have missed them.

Searching for Symptoms of Dementia

Suppose that you go to your physician complaining of, say, pain in the eye, or perhaps a skin rash. Your physician could look up every possible disease that could cause either of those symptoms. However, she is more likely to reach for French's *Index of Differential Diagnosis*, a text that practising physicians have relied on for years, and one that can still be found in every medical library. This popular reference book allows a physician to research a single symptom and find all the diseases in which that symptom may be present.

In the preface to French's first edition, published in 1912, the author writes, "The guiding principle throughout has been to suppose that a particular symptom attracts notice in a given case, and that the diagnosis has to be established by differentiating between various diseases to which this symptom may be due . . . but reference to the minor symptoms and physical signs which have not been thought sufficiently important to merit separate articles will be found in the general index at the end of the volume."[9]

Even as late as 1945, in French's *Index of Differential Diagnosis*, there is only one line stating that dementia and extreme memory loss are a symptom of senility associated with old age.[10] The most recent edition, published in 1996, devotes more than three pages to the subject.[11]

Current texts of internal medicine dedicate large amounts of space to the subject of dementias. A recent edition of *Scientific American Medicine*[12] dedicates an entire chapter to Alzheimer's disease and the dementias. About 80 per cent is devoted to Alzheimer's. Other dementias caused by prion diseases, and other neurologic disorders such as Parkinson's and Amyotrophic Lateral Sclerosis (ALS), are also mentioned. It is ironic that the dementia associated with syphilis – the only dementia worth noting in early texts – merits only a single line in the entire chapter.

Sigmund Freud's Observations

Perhaps early medical writings omit mention of dementia because it was a topic for psychiatry and not medicine. If so, the world's foremost psychiatrist, the founder of psychoanalysis, Dr. Sigmund Freud (1856–1939), must have dealt with it. Although the pro- and anti-Freud camps may never agree on the worth of his achievement, all will acknowledge that Freud explored the human psyche to depths never before plumbed.

In 1886, Freud set up a private practice in Vienna specializing in nervous disorders. At this time early in his career, he was recognized as an expert in cerebral palsy in children. His interest gradually shifted, and with his publication with Josef Breuer in 1895 of *Studies in Hysteria*, he moved from being a neurologist to being a psychiatrist. Freud founded the "science" of *psychoanalysis*, a term he introduced in 1896. Psychoanalysis emphasized the introspective study of the self.

During his life, Freud wrote voluminously on all aspects of human behaviour, both normal and abnormal. His theories of how the psyche develops and how mental diseases arise were so compelling that for the next three-quarters of a century almost the entire field of psychiatry relied on these theories for diagnosing and treating mental illnesses.

Curiously, although Freud was trained as a neurologist (one who deals with physical diseases of the central nervous system – the brain and spinal cord – and its peripheral nerves), the brain, as such, played little role in his theories. He attributed everything in the realm of mental illness to something that interfered with how the psyche (essentially, the mind) develops.

With the increasing acceptance of psychoanalysis, psychology and psychiatry turned away from looking for physical causes of mental illness and began looking for inner psychic disturbance and early childhood traumas. There was almost no suggestion that a physical illness in the brain could be responsible for a malady that manifested in what we would call a psychiatric disease. This contrasts sharply with our modern neurochemical model of psychiatric disease. As some psychiatrists have noted, "After Freud, psychiatry was brainless; in modern times it has become mindless."

Freud collaborated with some of the greatest medical minds of his time and produced voluminous works outlining his observation of all forms of psychopathology. He wrote a great deal on the subject of dementia, but only as it applied to those with schizophrenia – that is "*dementia praecox*." He did not speak of dementia occurring in any other framework. Although he must have been aware of the dementia associated with late-stage syphilis, he did not mention this disorder in his writings, perhaps because he categorized it as a medical and not a psychiatric illness.

Freud published twenty-four volumes of his observations and theories between 1895 and 1939. These volumes have been collected in *The Standard Edition of the Complete Psychological Works of Sigmund Freud*.[13]

The relevant index entries in this massive work list Delusions of persecution to *Dementia paranoides* (the dementia associated with advanced paranoia) to Demonic power and taboo. And there is nothing under memory loss or senile dementia either. Once again, we must ask, if Alzheimer's disease existed on any sort of scale in the early twentieth century, could Freud possibly have failed to note it?

Other Medical Observers

In the book *Three Hundred Years of Psychiatry 1535–1860*, authors Richard Hunter and Ida Macalpine describe early papers and books that mention senile dementia.[14] They note one case found in a book published in 1694, *Iatrica: seu, Praxis Medendi; The Practice of Curing, being a Medicinal History of many Famous Observations in the cure of Diseases, performed by the Author hereof, William Salmon.* The author of this early work, William Salmon (1644–1713), lays out his affidavits nicely in his subtitle:

> Being a medicinal history of above three thousand famous observations in the cure of diseases, performed by the author hereof: together with several of the choicest observations of other famous men . . . : wherein for the most part you will find 1. the constitution of the body of the sick, 2. the symptoms predominant, 3. the cause of the disease, what? 4. the exact method which was taken in the cure, 5. an exact account of the medicines exhibited, with the order of their exhibition, various doses and success thereupon . . . / perform'd by William Salmon.

Clearly a keen observer and medical practitioner, Salmon describes but one case of dementia.

Another oblique reference to memory fading with age is described in *The Art of Memory: A treatise useful for such as are to speak in publick*, written in 1697 by Marius D'Assigny (1643–1717). As to the cause of memory loss, D'Assigny notes the adverse effects of "the ill fumes of the stomach" ascending to the brain to memory's impairment. He devotes a full chapter to aids in assisting the memory, including liniments, ointment, sneezing powders, and plasters, all of which he recommends to prevent the impairment of the memory by such fumes. He suggests an experiment:

> Take the seed of Orminium, and reduce it to Powder, and every Morning take a small quantity in a Glass of Wine. And

they say that the Shavings or Powder of Ivory produce the same Effect, namely, the corroborating of the Brain and Memory; as likewise a Grain of white Frankincense taken in a Draught of Liquor when we go to Bed, dries up the offensive Humors of the Brain. And it hath been observed, that the Application of Gold to that Sutura which divides the Seat of Memory from the other Closets of the Brain, strengthens the Weakness of the Head, drives away all Pain, and hath a wonderful Effect upon the Faculty of Memory.[15]

Would that our modern pharmacies carried such a glorious array of remedies. No doubt some of the modern equivalents found bottled on the shelves of drugstores and health-food stores will prove as efficacious for our brains as a Grain of white Frankincense.

The ensuing centuries yield a few more references to dementia, but the overwhelming impression is that although dementia associated with aging did exist, it was extremely rare. A similar review of other books on neurology and psychiatry from the late nineteenth to early twentieth century reveals no mention of anything that resembles AD.

In 1888, the recognized standard text of neurology was the two-volume *A Manual of Diseases of the Nervous System*[16] by British neurologist Sir William Richard Gowers (1845–1915). Gowers, a talented painter and etcher, whose paintings were shown by the Royal Academy of Arts, possessed extraordinary powers of observation. His masterwork, often called the Bible of neurology, is a classic noted for the remarkable illustrations done by his own hand. Once again, we find that this book has listings for neither dementia nor senile dementia.

Medical Journals

The earliest medical journals appeared in Europe in the seventeenth and eighteenth centuries, but did not last for long. The beginning of the nineteenth century ushered in a new age in the dissemination of medical information with the emergence of the general medical journal. In England, the *Lancet*, a publication independent of any medical society, was first published in 1823, and the *British Medical Journal*, the journal of the British Medical Association (BMA), appeared in 1857. Now physicians could report advances or interesting cases without having to wait for the publication of the next edition of a standard text. The lag time between the discovery and publication was reduced from several years to several months. (Today, the Internet has again reduced this time to weeks or even days.)

But in order for physicians to find articles of interest to them, they needed an annual indexed listing of all medical articles published each year. John Shaw Billings recognized that need.

In 1865, John Shaw Billings (1838–1913), a young army surgeon just discharged from his surgical duties in the U.S. Civil War, and not yet thirty years old, took control of the National Library of Medicine in the United States. By 1876, he had increased the modest collection of medical books and journals to forty thousand volumes and collected a similar number of pamphlets. By 1880, he published the first series of the great medical *Index Catalogue* of the Library of the Surgeon General's Office, which became his life's work. Thirty years later when Billings retired from the National Library of Medicine, he left behind the largest medical library in the world.

Simultaneously with the huge catalogue he published the *Index Medicus*, initially a quarterly and later, as the volume of material increased, a monthly bibliography of medical literature. The National Library of Medicine still publishes the monthly *Index* today.

Reviewing these indexes for evidence of dementia yields some interesting results. Until 1921, we find no citation for the subject of dementia, and only a few each year for senile insanity. These references do not relate to anything resembling Alzheimer's disease. For example, a 1903 article is optimistically titled "Home Treatment of Senile Insanity." Along with that implausible title, we also find "Acute Dementia in an Old Man; Recovery After Two and One Half Months of Treatment." In 1921, one article discussed the relationship between recurrent dreams and senile dementia. Only in the 1930s did articles on what may have been Alzheimer's began to appear. By contrast, the year 2002 alone lists 2,897 articles on dementia and 1,903 articles on Alzheimer's disease.

Popular Medical Literature

The popular medical literature of the early twentieth century, *The Home Health Manuals* and books of popular remedies, are similarly devoid of any mention of Alzheimer's or anything resembling it, except for the form of dementia that occurs in late-stage syphilis.

Home health guides, very popular in the first half of the twentieth century, remain big sellers today. Home Health Society of London, Philadelphia, and New York published the 1914 edition of *Health and Longevity*. Twenty physicians from all over the world edited this text. The cover page immodestly states it is the "Absolute Authority on Every Subject" and "No other medical work in the world has such a list of eminent physicians and specialists for every subject."[17]

At more than thirteen hundred pages long, the book features excellent sections on anatomy and good descriptions of most diseases classified by symptoms. The only mention of dementia is in one sentence as a "natural termination of extreme old age." There is nothing of any of the other neurodegenerative diseases. Dementia is listed along with Melancholia, Imbecility, and Idiocy in the section on "Insanity and Madness."

Another popular book was *The People's Common Sense Medical Adviser in Plain English, or Medicine Simplified.*[18] The face sheet declares: ONE HUNDREDTH EDITION, Four Million Five Hundred Thousand. If we believe those figures, this must have been one of the most popular home medical guides ever written. In the medical dictionary that is part of this book, there is no listing for dementia or senile dementia, nor are any similar conditions listed or discussed in the parts of the book dealing with old age.

Another standard home reference book, *Pannell's Reference Book for Home and Office, 1906*, contains, among other things, an English dictionary and a medical dictionary. It contains no listings relating to dementia or senility.

CHAPTER

OBSERVATIONS ON THE ELDERLY: ANCIENT GREECE TO MODERN LITERATURE

If medical observers did not notice dementia in the past, perhaps literature will give a more rounded picture of the prevailing opinions and assumptions about the elderly through the ages.

SMELLING LIKE A GOAT:
THE ELDERLY IN CLASSICAL GREECE AND ROME

Literature and mythology from classical times universally portray old people as being in full command of their mental faculties. The singular, most readily noticeable characteristic of Alzheimer's victims is memory loss. In ancient Greece, even the eldest of the poets – far from losing their memory – could recite thousands of lines from memory. In Greek literature, although the aged could be portrayed as vain, foolish, and occasionally gullible, there is no character in this entire body of literature who could be said to be senile.

Mental illness did, of course, exist. Romans prepared wine by simmering a syrup over a fire. Unfortunately, their cooking vessels were made partly or wholly of lead. Little did the Romans know that lead poisoning could lead to mood swings, irritability, severe abdominal pain, headaches, and loss of motor coordination, among other symptoms.

However, medical writers of the time dismissed poisoning as a cause of mental illness, believing rather in a strictly physiological cause. For such affliction, doctors prescribed a remarkable battery of treatments: ". . . cuppings, pummelings, massages, purgatives, and diets. (Fortunately for them only a few victims could afford the cost of treatment.)"[1]

Researchers encounter great difficulty trying to determine accurate demographics for ancient civilizations. Funerary inscriptions, while notoriously suspect, nevertheless provide some indication of the proportion of old people in the population. A classics scholar, B.E. Richardson, collected 2,022 funerary inscriptions from the classical period in ancient Greece. Of the 2,022 deceased, 3.8 per cent died between the ages of 66 and 75 years, 2.6 per cent between 76 and 85 years, almost 1 per cent between 86 and 95 years, and 9 of the 2,022 people (0.45 per cent) died between the ages of 96 and 110.[2]

If the current rates of Alzheimer's disease (one in ten at age sixty-five, and doubling every decade after) were present during this ancient period, we would expect about thirty of these people (about 1.5 per cent of the population) to have had Alzheimer's disease when they died. Yet as we have seen, such was not the case.

The same B.E. Richardson also compiled the ages of death of forty-eight eminent Greek philosophers. Of this group only one died before age sixty, five died in their sixties, nine in their seventies, nine in their eighties, seven in their nineties, and the remainder were said to have died at an old or extremely old age. This is a group of men whose lives are well chronicled – and there is absolutely no evidence that any one of them suffered from anything that resembled Alzheimer's disease. An examination of the records of the ancient Roman world reveals similar figures.

The late Sir M.I. Finley, professor emeritus of ancient history at Cambridge, noted that he had great difficulty finding any reference to mental illness in old age in ancient civilizations.

> In the Hippocratic *Aphorisms*, there is a brief section enumerating the diseases characteristic of each age-group; the list for the old includes catarrhal coughs, arthritis, nephritis, apoplexy [i.e., stroke], insomnia, failing sight, deafness, and a few others, *but not dementia or anything else that they (or we)*

would recognize as mental illness [our emphasis]. I do not conclude from this silence that senile dementia was non-existent in antiquity, but I know of no way to penetrate the almost total silence.[3]

The Greeks did not ignore or hide dementia because of some idealized concept of old age. They were enthralled by physical beauty, and feared and reviled the idea of old age. They simply did not discuss dementia because it did not exist to any noticeable extent.

The late professor Maria Haynes, known for her study of the concept of old age during the Greek and Roman empires and the Middle Ages, compiled a list of all the terms used to describe the elderly in ancient literature. The list from the literature of republican Rome includes, in order of frequency: "dirty; sallow complexion; stinking breath, smelling like a goat; . . . untidy; . . . groaning and of damnable shape . . . ugly old thing; a withered, worn-out, flabby old man; the old fossil; decrepit old frame, and stupid old chatterbox."[4] Dementia is not mentioned.

Aristotle wrote a scathing diatribe against the elderly, accusing them of every sort of physical, moral, and mental failing – with the exception of dementia. He goes on for pages about the faults of old men, calling them "timorous, hesitant, suspicious, parsimonious, fearful, cowardly, selfish, pessimistic, talkative, avaricious and ill-humoured."[5]

Here is but a short sample of Aristotle's rant against old men, which runs for several thousand words.

> They think evil; that is, they are disposed to put the worse construction on everything. Further, they are suspicious because they are distrustful, and distrustful from sad experience. As a result, they have no strong likings or hates; rather, illustrating the precept of Bias, they love as men ready some day to hate, and hate as ready to love. They are mean-souled, because they have been humbled by life. Thus they aspire to nothing great or exalted, but crave the mere necessities and comforts of existence. And they are not generous. . . . They are cowards, apprehensive about everything – in temperament just the opposite of youth.[6]

Yet nowhere in this tirade does Aristotle suggest any kind of mental derangement. Certainly he would not have failed to use dementia against the elderly if such a trait were common.

As for the Romans, neither did they see mental decline as a characteristic of old age. Cicero (106–43 B.C.E.) scorns the need for physical adroitness, and notes that the elderly achieve great things through mental capabilities:

> It is not by muscle, speed, or physical dexterity that great things are achieved, but by reflection, force of character, and intellect; in these qualities old age is usually not only not poorer but is even richer.

However, the Romans did not generally glorify old age. As with the Greeks, the opposite was often the case. But in none of the often-vitriolic descriptions of the aged is there any mention of what we would consider dementia.

We can only conclude that Alzheimer's disease did not exist during the classical ages of Greece or Rome or, if it did, was so rare as to escape notice.

HIS EYE WAS NOT DIM:
THE ELDERLY IN RELIGIOUS WRITINGS

Members of the Christian Church (along with other community and cultural groups and native people) to this day refer to the wise people who govern the church as the elders. Biblical texts speak of the sages, the patriarchs, and the wise elders. In Biblical times, old age was considered a reward for a good life. All the Biblical patriarchs lived to an old age and were considered wise till the day they died. The Bible is full of references to old age:

The old as wise:

With the ancient is wisdom; and in the length of days understanding.

– Job 12:12

Strength in old age:

And now, behold, the Lord hath kept me alive, as he said, these forty and
five years . . . and now, lo, I am this day 85 years old. And I am yet as
strong this day . . . as my strength was then, even so is my strength now.

– Joshua 14:10-11

The health of the elderly:

And Moses was a hundred and twenty years old when he died: his eye
was not dim, nor his natural force abated.   – Deuteronomy 34:7

Old age as a reward:

   Old age was seen as a reward from God for a lifetime of obedience to
the commandments:

And if thou wilt walk in my ways, to keep my statutes and commandments . . . then I will lengthen thy days.   – 1 Kings 3:14

Elders as a resource:

   The Bible also refers to the aged as a resource for the community:

And Moses was 80 years old, and Aaron 83 years old, when they spake
unto Pharaoh.   – Exodus 7:7

The aged women likewise, that they be . . . teachers of good things.

– Titus 2:3

Jewish Religious Texts

In Jewish religious texts, one section is called the Wisdom of the Fathers.
The commentaries on these texts note that the phrases *old man* and
scholar are to be considered interchangeable.

Maimonides, a revered rabbi and commentator on the Torah, discusses a person's duty to study.

> At five [one begins the study of] the Bible. At ten the *Mishnah* [commentaries on the Bible]. . . . At fifteen the Talmud [also commentaries on the Bible]. At eighteen marriage. At twenty to pursue a livelihood. At thirty strength. . . . At fifty, one gives counsel. At sixty [one reaches] old age. At seventy [one reaches] the fullness of age. At eighty [one reaches] strong old age. At ninety [one is] bent. And at one hundred it is as if one had already died and passed from the world.
> – Pirkei Avot (The Wisdom of the Fathers) 5:21

Thus we can find no evidence in any of the western religious cannon of anything resembling Alzheimer's.

BRITISH LITERATURE

The consummate literary icon for derangement, Shakespeare's King Lear, stands out as an extraordinary and memorable portrait of madness. Countless scholarly texts have discussed endlessly the nature of Lear's insanity. But the text makes clear that whatever tormented King Lear, it was not Alzheimer's disease. His madness was of relatively sudden onset, his memory remained intact, and he remained lucid, his linguistic faculties undiminished. In fact, Shakespeare may have used insanity as a metaphor for Lear's changing perception of himself and the world in which he lived. In his madness Lear gains new insight into relationships, loyalty, selfishness, justice, dignity, and poverty and wealth. An Alzheimer's victim, on the contrary, loses all sense of meaning of any abstract concepts.

Other characters more fully reveal the perception of the aged during the late Renaissance period. Lear's daughter Goneril comments, "As you are old and reverend, you should be wise." In the same play, the Fool says, "Thou should'st not have been old till thou hadst been wise." Shakespeare and his audiences clearly share the expectation that old age normally brings wisdom not confusion.

Sir Francis Bacon, a contemporary of Shakespeare, wrote a lengthy comparison of youth and age in 1620. Like the classicists of Greece and Rome, Bacon finds much to criticize in the elderly.

> Youth has modesty and a sense of shame, old age is somewhat hardened; a young man has kindness and mercy, an old man has become pitiless and callous; youth has a praiseworthy emulation, old age an ill-natured envy; youth is inclined to religion and devotion by reason of its fervency and inexperience of evil, in old age piety cools through the lukewarmness of charity and long intercourse with evil, together with the difficulty of believing.
>
> . . . Youth is liberal, generous, and philanthropic, old age is covetous, wise for itself, and self-seeking; youth is confident and hopeful, old age diffident and distrustful; a young man is easy and obliging, an old man churlish and peevish; youth is frank and sincere, old age cautious and reserved; youth desires great things, old age regards those that are necessary; a young man thinks well of the present, an old man prefers the past; a young man reverences his superiors, an old man finds out their faults; and there are many other distinctions which belong rather to manners than the present inquiry. Nevertheless as old men in some respects improve in their bodies, so also in their minds, unless they are quite worn out.[7]

Notice that Bacon only hints at the slight possibility of the aged experiencing a decline in mental power, and only if "they are quite worn out."

Alfred Lord Tennyson wrote "Ulysses," perhaps the best-known poem dealing with aging, in 1842, when he was just thirty-three years old. The aging Ulysses says:

> Tho' much is taken, much abides; and tho'
> We are not now that strength which in old days
> Moved earth and heaven; that which we are, we are, –
> One equal temper of heroic hearts,

Made weak by time and fate; but strong in will
To strive, to seek, to find, and not to yield.

Hardly the sentiment of one who fears the onset of Alzheimer's.

Robert Browning's poem "Rabbi Ben Ezra" (1864) gives us the famous line, "Grow old along with me! / The best is yet to be." With our modern fear of dementia, it is difficult to imagine any contemporary poet writing those lines.

One notable exception to the dearth of demented elders is found in the writing of British author Jonathan Swift. Swift wrote of the Struldbrugs, a race whose members live forever but lose their mental faculties in their old age.

> At ninety, they lose their teeth and hair; they have at that age no distinction of taste, but eat and drink whatever they can get, without relish or appetite. The diseases they were subject to still continue, without increasing or diminishing. In talking, they forget the common appellation of things, and the names of persons, even of those who are their nearest friends and relations. For the same reason, they never can amuse themselves with reading, because their memory will not serve to carry them from the beginning of a sentence to the end; and by this defect, they are deprived of the only entertainment whereof they might otherwise be capable.
>
> – Jonathan Swift, *Gulliver's Travels* (1726)

Swift had seen his uncle decline mentally, and in an ironic twist of fate, the same thing happened to him. Swift's description of dementia endures as one of the very few in the English literature to this point. The myth persists that Swift was driven insane by his misanthropy, but modern medical opinion suggests he probably suffered from Ménière's disease, an affliction of the inner ear that causes imbalance and dizziness, and may have caused his erratic behaviour in later life. He also suffered a stroke, which may have magnified his symptoms. But he seems to have kept his sense of humour. At his death in 1745 he left money to establish St. Patrick's Hospital for "ideots and lunaticks" because "no nation wanted [i.e., needed] it so much."

FRENCH LITERATURE

One of the oldest works of French literature, the "Song of Roland" (*circa* 1200), is the first of the great French heroic poems known as *chansons de geste*. An epic tale recounting the heroic exploits of Charlemagne, the poem tells us of a lengthy and bloody battle between the French and the Spanish.

Near the end of the poem, when Charlemagne is old and tired, he wishes only to die. The archangel Gabriel appears to Charlemagne in a dream, and tells him that God will not let him retire. Pagans have besieged a citadel, and Charlemagne must depart to fight yet another battle.

> Right loath to go, that Emperor was he:
> "God!" said the King: "My life is hard indeed!"
> Tears filled his eyes, he tore his snowy beard.

Weary and weeping, but dutiful to God, the old and tired Charlemagne prepares for another bloody clash – a picture of the aged warrior far removed from our view of the elderly wasting away in nursing homes today.

The great French novelist Victor Hugo wrote a playful poem about aging in 1875, when he was seventy-three years old. In it, he admonishes his grandchildren:

> Ah! Do not elevate me to the rank of God!
> You see, I'd do all sorts of strange things,
> I would laugh, I would take pity on roses,
> On women, on victims, on the weak and the trembling; . . .
> I'd be a good guy if I were the Good Lord.
> > – Victor Hugo, *"L'Art d'être grand-père"* (1877).
> > Translation by P. Archambault

André Gide writes about sexual desire that persisted into old age.

> Even at the age of eighty, things like this are not easy to
> admit. King David was just my age, no doubt, when he used to
> invite the young girl Abishag to come and warm up his bed.[8]

Again we do not find an idealized portrait of the aged. In her last
autobiographical volume written when she was sixty-four, Simone de
Beauvoir writes, "Society looks upon old age as a kind of shameful secret
that is unseemly to mention."[9]

Although the French did not glorify old age, almost nothing in any
historical French literature portrays the elderly as demented. Once
again, Alzheimer's did not seem to exist. A review of other European literature yields similar results.

ASIAN LITERATURE

Japan has the longest life expectancy of any country in the world (76.36
years for men and 82.84 years for women). The Japanese respect the
elderly, defer to them, and even honour them each September with a
national holiday called *Keiro no hi*, or Respect for the Aged Day.

In Japanese folk tales, the elderly are portrayed as wise. Of the Seven
Gods of Fortune, five are depicted as old men.[10]

The fourteenth-century Buddhist priest Kenko tells us that the longer
a man lives, the more shame he endures. Yet he does not suggest the elderly lose their faculties, only their sensitivity:

> In his sunset years, he dotes on his grandchildren, and prays
> for a long life so that he may see them prosper. His preoccu-
> pation with worldly desires grows ever deeper, and gradually
> he loses all sensitivity to the beauty of things, a lamentable
> state of affairs.[11]

A more modern writer, the prolific Japanese novelist Junichiro Tanizaki
(1886–1965) wrote *Diary of a Mad Old Man*, which is the journal of
Utsugi, a seventy-seven-year-old man of refined tastes who is recovering

from a stroke. Utsugi finds that even though his body is falling apart, his carnal appetite continues unabated (a not uncommon theme among the elderly). He tells us, with wry good humour, his thoughts on aging and sex.

> I know very well that I am an ugly wrinkled old man. . . . How could anyone with a face like this ever hope to appeal to a woman? Still, there is a certain advantage in the fact that it puts people off guard. . . . And to make up for my own inability, I can get her involved with a handsome man, plunge the whole house into turmoil, and take pleasure in *that*.[12]

Early Chinese writings are dominated by the influence of Confucius, who saw one of his missions as the upholder of the best in ancient Chinese traditions. One of the most important of these traditions was the sincere practice of the rituals and customs calling for consideration and respect for the elderly, especially one's parents.[13]

The Twenty-four Examples of Filial Piety (*Er-shih-szu hsiao*) consists of brief didactic anecdotes first compiled by Kuo Chü-ching in the fourteenth century. Various other versions followed. Even today, these stories form an important part of Chinese folklore. One of the Examples tells the story of Lao Lai-tzu.

> He provided lavishly for his father and mother. Although he was seventy years of age, he would not concede that he was old . . . He would also carry water into their front hall, and then pretend to be childishly awkward and fall to the ground, in this way amusing his parents and making them happy.[14]

In China, even today, both men and women older than seventy discuss their age with pride. In some colloquial versions of this story, Lao Lai-tzu does not mention his age to protect his parents from the realization that not only they, but also their son, might be near death.

Both traditional and modern Chinese writing explore the theme of aging parents and their relationship to their families. In no case are these aging people portrayed as anything but alert and completely aware of everything that is going on. Nothing in this literature bears even the slightest resemblance to Alzheimer's disease.

WAS DEMENTIA BENEATH NOTICE?

Perhaps Alzheimer's did exist and people just thought it was part of the normal aging process. Perhaps it was so common that it was felt to be obvious, and since everyone was aware of it there was no point in writing about it.

If that were the case, we would expect that other common features of old age would be equally unworthy of mention. However, we find ample discussion of all the characteristics, distinctions, peculiarities, and eccentricities of old age in mythological, religious, and lay literature.

Two themes emerge in looking at records of the elderly in previous centuries:

- The elderly are regarded as wise, and worthy of reverence and honour.
- The elderly are regarded as physically debilitated, with cranky, irritating, or deficient personalities.

Neither of these characterizations implies dementia.

In ancient literature, mythology, and religious testaments, we found frequent mention of old people being physically weak, shrunken, and wrinkled, but seldom, if ever, demented. Nowhere in the ancient literature are the elderly considered either forgetful or senile.

We do have a few anecdotal reports of dementias from the eighteenth century, although nothing even approaching the scale that would be comparable to today's prevalence. Given the descriptions, these dementias may have been vascular or may have been the result of other diseases such as syphilis or Ménière's disease.

Even though a few demented people existed in centuries past, the concept of the elderly as *inevitably* dotty is a very modern one indeed.

Numerous old adages show respect for the wisdom of the elderly, such as the African proverb: "When an old man dies, a library burns down." Just half a century ago, Harvey C. Lehman compared young and old, and came out in favour of the old.

> ". . . [T]he old usually possess greater wisdom and erudition. These are invaluable assets. To learn a new [thing] they often have to unlearn the old, and that is twice as hard as learning without unlearning. But when a situation requires a store of past knowledge then the old find their advantage over the young."[15]

In *History of Old Age*, a comprehensive and scholarly review of old age from prehistoric times to the present, Georges Minois talks at length about the quickly growing number of elderly in our society. But he does not once mention Alzheimer's disease or dementia, even though he wrote the book less than two decades ago.

More recently, with burgeoning numbers of AD victims, and as baby boomers find old age looming nearer, acerbic references to "old-timer's disease" (as a play on Alzheimer's disease) have become common. The first known citation of the half-joking phrase *senior moment* occurred less than a decade ago in the *Daily News* (New York).

> Evelyn Weinstein, who works with 35 nursing homes in Nassau County, came to the conference for herself and her patients. "I really want to learn what I can for our people," she said. "And I've had a few 'senior moments' myself."[16]

As the epidemic of AD grows ever larger, the forgetful senior has become a stereotype. *Webster's New World College Dictionary* named the phrase *senior moment* the Word of the Year 2000.

Even the words describing senile dementia are of modern origin. Strictly speaking, the term *senile* simply means *old*. So medical science uses the term *pre-senile dementia* to describe Alzheimer's disease that develops before age sixty-five, while *senile dementia* refers to AD that develops after age sixty-five (even though it's the same disease). In common lay usage, however, the word *senile* implies a loss of mental faculties.

The Complete Oxford Dictionary, in addition to defining words, also cites the first time any word is used in print. The word *dementia* is first used in 1874. The word *senile*, meaning belonging to, suited for, or incident to old age, dates back to 1661. But *senile*, referring to the *weakness* of old age, first appears in 1848.

More telling, senile meaning some form of *mental* condition does not appear until 1962. The citation reads, "1962 *Lancet* 8 Dec.: 1212–3. Of every 100 potential long stay male patients, 30 were schizophrenics, 21 seniles, and 12 manic depressives."

It is not often that you can find almost exactly when a new word or a new meaning for an old word enters the language. In the 1961 edition of the venerable reference book *Roget's Thesaurus of the English Language*, the entry for the word *senile* reads:

"*adj.* decrepit, infirm, anile (oldness, weakness)."

The entry for *senility* reads:

"*noun.* dotage, caducity, second childhood, decrepitude (oldness, weakness)."[17]

Just one year later, the 1962 edition of *Roget's International Thesaurus* shows this entry for *senile*:

"*n.* dotard, and *adj.* Unperceptiveness, incomprehension, blindness, short-sightedness, stupidity, dumbness, doltishness, asininity, density, dullness, obtuseness, dull-witted, thick-wittedness."

The 1962 entry for *senility* reads:

"*n.* senile dementia, feeble-mindedness, and dementia."[18]

In the space of one year the popular meaning for senility changed from old to feeble-minded and demented – showing a significant change in our perception of the elderly, and further evidence for the marked increase in AD in our society.

MODERN LITERATURE

As AD has become more prevalent, so have references to it become common. Modern literature has given us the darkly comic *Barney's Version* (1997) by Canadian novelist Mordecai Richler. Barney worries about AD, but doesn't realize he actually has the disease. The novel is a first-person "memoir" full of mistakes (corrected in footnotes written by Barney's son), and a loving portrait of a curmudgeonly character with an intellect in decline. In the non-fiction genre, John Bayley wrote *Elegy for Iris* (U.S. version), also known as *Iris: a Memoir of Iris Murdoch* (U.K.) in 1999 while his wife, Dame Iris Murdoch, a pre-eminent novelist of the twentieth century, was still living with the disease.

These and a growing body of other works accompany the rise in AD in recent years, in stark contrast to the almost complete silence on the topic in years gone by. It's hard to avoid the conclusion that Alzheimer's is, in effect, a new disease.

CHAPTER

AGE PATTERNS AND ALZHEIMER'S DISEASE

We have seen that Alzheimer's disease is now so common an affliction of elderly people in developed western nations that it can reasonably be described as an epidemic. And yet, it appears that the disease did not exist, or occurred only in exceedingly rare instances, before the beginning of the twentieth century. There is no mention of any condition resembling Alzheimer's in any of the most authoritative medical reference books until some years after it was first described, in 1906. And a survey of the literature of several nations supports the idea that old age, before the middle of the twentieth century, was associated with decrepitude and physical decline – but almost never with dementia. This finding seems, on the face of it, remarkable.

But is there another reason why Alzheimer's went unnoticed until comparatively recently? There may be an answer in terms of population patterns: Perhaps, because Alzheimer's is a disease of aging, it may have been very rare because people did not live as long in past centuries as they do today. Maybe people simply did not live long enough to contract the disease.

ARE WE REALLY LIVING LONGER NOW?

People on average are, of course, living longer than they used to centuries ago. However, most of these gains have been made at the bottom end of the scale.

In ancient Rome, more than a quarter of all live-born babies died before their first birthday. About one-third of the children who survived infancy were dead by the age of ten.[1]

The average life expectancy in Europe in the sixteenth to seventeenth centuries was only about thirty-five years, mostly because so many people died very young. In eighteenth-century Europe, rates of infant and childhood (up to age five) mortality were about 212 and 177 respectively. This means that of 1,000 children born, 389 (i.e., 212+177) died before their fifth birthday.[2] Figuring these early deaths into the overall death rates makes the average life expectancy very low.

Age Pyramids

To get a good idea of the age/sex distribution population of a particular place, we can diagram it on an age pyramid. An age pyramid reveals in graphic form the ages of the population of a particular place at a given time.

These diagrams show how many people of every age that were alive in one place, France, at different moments in time – 1790, 1911[3], and 1995. The left side shows how many males were alive, while the right side shows the number of females. The column in the centre shows the ages of the people alive.

If you look at age forty in the central column, you can see that the bars for both men and women are only half as long as the bars at the base of the pyramid. This shows us that in 1790 more than half of the population had died by age forty. Looking farther up the column, we can see that by age sixty, more than 75 per cent had died.

This pyramid shape is typical of most populations in the past, and of less-developed nations today – very high birth rates produce a broad base of infants and children, bars that shrink rapidly with increasing age due to high death rates.

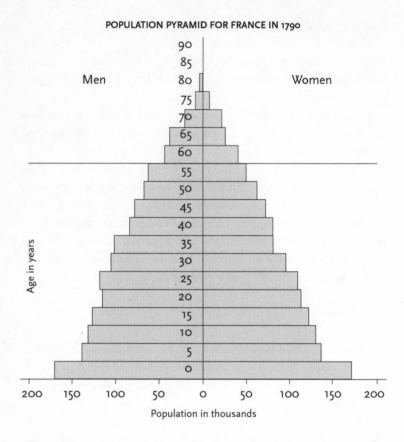

POPULATION PYRAMID FOR FRANCE IN 1790

Men — Women

Age in years

Population in thousands

Skip ahead a century or so, and the average life expectancy has increased greatly. The age pyramid is now shaped like a beehive.

In 1911 in France, more than half the people were still alive by age sixty – showing an increase in life expectancy, but with all the gains at the bottom of the pyramid.

Moving into the end of the twentieth century, we see a completely different shape of age pyramid. More developed nations today tend to have a shape more like a silo – low birth rates (small base) and low death rates (everyone survives into old age). By 1995, almost 65 per cent of the population of France were still living at age sixty.

The more interesting comparison emerges if you cut off the three pyramids at age sixty. The shape below the cut varies markedly in the

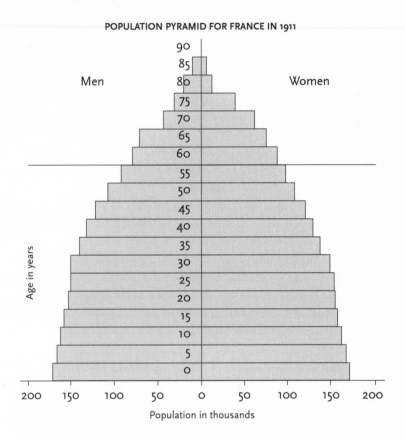

POPULATION PYRAMID FOR FRANCE IN 1911

Men   Women

Age in years

200   150   100   50   0   50   100   150   200

Population in thousands

three diagrams. In 1790, the shape of the lower part of the pyramid has very steeply sloping sides, indicating a significant drop in the population every decade until age sixty. By 1911, the slope below age sixty is greatly decreased, indicating fewer people died in the years prior to age sixty. The 1995 pyramid shows very little slope below age sixty, indicating that currently almost everyone lives to greater than age sixty. (The bulge in the middle, the so-called pig-in-the-python, represents a period of official incentives for larger families.)

However, the section of the pyramid above age sixty looks very similar for all three years. In other words, the proportion of people older than the age of sixty remains the same over several centuries, decreasing by about 25 per cent every decade after sixty.

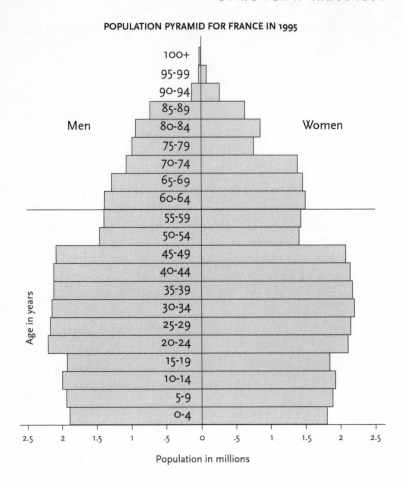

POPULATION PYRAMID FOR FRANCE IN 1995

Men   Women

Age in years

2.5   2   1.5   1   .5   0   .5   1   1.5   2   2.5

Population in millions

Curiously, the percentage of seniors in a given population doesn't seem to change much over the centuries. In the population pyramids, the ratio of elderly, i.e., those older than sixty, to the twenty to sixty group remains about the same in all the pyramids. In 2001 in both the United States and Canada, more than 12 per cent of the population, or nearly 4 million Canadians and about 34 million Americans, were sixty-five or older.[4] In the United Kingdom, more than 5.3 million citizens, 14 per cent of the population, is sixty-five or older today.

This comparison clearly shows that not only have there always been elderly people in our society, but the ratio of ages of those older than sixty has remained pretty much unchanged for at least the last three hundred years.[5]

Census Records

We can also examine census records for various years to see how many old people there are at any one time. According to the U.S. census figures, if you were born in 1950 you would have a life expectancy of 68.2 years. By the year 2000, this had increased to 76.9 years – a gain of 8.5 years. However, if you were 65 years old in 1950, you could anticipate living for 13.9 more years. By 2000, if you were 65, you would expect to live for another 17.9 years – only 4 years longer than in 1950. This again illustrates that even in modern times the gains in life expectancy made over the age of 65 have been quite modest.

In fact, a few centuries ago, the rate of childhood death was so high in certain cultures that children were not even named until they reached a year of age. Thus perhaps parents hoped to lessen the impact of their dying. Women had many children in the hopes that one or two would live to adulthood. With the advent of antibiotics, modern childbirth methods, and immunization, almost all children in developed countries today can look forward to reaching adulthood. Childhood is where most of the gains in life expectancy have been made.

The accuracy of statistics before 1800 varies widely; however, some English church parishes kept good records, and there are good civilian records for the Scandinavian countries going back to about 1750.[6]

Examination of the data from Scandinavian countries reveals some interesting figures. A married woman would bear, on average, almost nine children (8.99). However, only about 2.3 to 2.8 of these children would live long enough to have families of their own. Thus the population of a village often remained the same for hundreds of years. Lack of means of birth control combined with a high rate of early mortality ensured a fairly stable population base.

If a male escaped all the dangers of simply being born, avoided the diseases of early childhood, and managed to attain the age of twenty-one, he could count on a reasonably lengthy life. His chances of reaching seventy or beyond were not greatly different than they are today.

The aristocracy of England may, in past centuries, have enjoyed an advantage over the common folk in their life expectancy, simply because the lower classes suffered more through poverty, injury, childbirth, and other life-threatening conditions. But even as long as five centuries ago, a twenty-one-year-old male could reasonably expect to live to age seventy-one and beyond.

EXPECTATIONS OF LIFE AMONG THE ARISTOCRACY OF ENGLAND, 1200-1745[7]			
Period	Number of males observed	Estimated number of years of life after 21	Age at death
1200–1300	7	43.14	64.14
1300–1400	9	24.44	45.44*
1400–1500	23	48.11	69.11
1500–1550	52	50.27	71.27
1550–1600	100	47.25	68.25
1600–1650	192	42.95	63.95
1650–1700	346	41.40	62.40
1700–1745	812	43.13	64.13

* The low number in the period 1300–1400 reflects the impact of the Black Death.

In France in 1740, the probability of a twenty-one-year-old reaching the age of sixty was 41.9 per cent for men and 43 per cent for women. By 1820, this had risen to 58.1 per cent for men and 59 per cent for women. In England in 1881, 6.9 per cent of men and 7.8 per cent of women were older than the age of sixty.[8]

In 1750, the risk of dying between the ages of five and ten years was about 150 per 1,000. By 1914, this risk had dropped to fifteen per thousand, and between 1963–1985, it was less than two. Similar gains were made in the infant-to-five and ten-to-fifteen age groups.

These figures clearly show that in the last three hundred years, although the average life expectancy has risen very sharply, the top of the age pyramid – that is the percentages of people who die in their sixties, seventies, and eighties – has remained almost unchanged.

Hundreds of cases document this information. Giorgio Vasari compiled a list of famous Italians in his *The Lives of the Most Excellent Painters, Sculptors and Architects*, first published in 1550 in Florence, and revised and extended for a second edition in 1568. Vasari lists forty-seven Italian artists from the fourteenth to the sixteenth century.

- thirty-four (or 72 per cent) lived to be older than sixty;
- seventeen (36 per cent) lived to be older than seventy; and
- six (12 per cent) were older than eighty when they died.

A census of serfs in a French abbey's domain in the ninth century shows that "even among this category of poor peasants, more than 11 percent of the adults were aged over 60."[9] Church records from Bristol, England, from the sixteenth and seventeenth centuries listing the age of death of all parishioners show that about 20 per cent died between the ages of sixty and sixty-nine and about 15 per cent died older than the age of seventy.

Records from England show the age of retirement and death of holders of one hundred of the highest administrative posts in the Tudor government from 1485 to 1558; most retired at about age sixty-three to sixty-five, and died about five years later. Not much different from what we might expect from senior civil servants today.

Public-Health Statistics

Finding little evidence for dementia of Alzheimer's in early literature, we also looked for evidence of AD in hospital and public-health statistics. In 1890, the population of the United States was 63 million, of whom 2.4 million were sixty-five or older. If the incidence of the disease were similar to today, there should have been about 240,000 cases of Alzheimer's or senile dementia. If these huge numbers existed, there should be some mention of people exhibiting these very distinctive symptoms. Yet, as we have seen, there is none.

The Virginia State Assembly for the Insane shows records for 754 patients from the period of 1868 to 1879. The diagnosis is listed for all the 754 patients. Only one case lists a diagnosis of "insanity of old age," which may correspond to a dementia. This is a rare example of age-related mental disease that can be found from that period.

Even as late as 1982, we find AD at a much lower rate than today. In 1982 in the United States, there were 31.8 million people older than the age of sixty-five. If today's rates of AD were prevalent in 1982, we would expect there to be 5.2 million of these seniors with the disease. Yet figures published at that time show that only 650,000 people suffered from dementia. This is only one-eighth of the expected number if rates of the disease were the same as today. Is it conceivable that seven-eighths of all people with AD went unnoticed? Again the figures point to the tremendous increase in the scope of the AD epidemic in a short period of time.

So although fewer people survived to live to a relatively old age in years gone by, many of those who survived childhood lived nearly as long as we do today. Although in absolute numbers there are certainly more elderly people today than at any other time in history, the same can also be said of young people. The point is that among the adult population, the number of elderly people compared with other adults has not changed in any significant way. Just like the poor, the elderly have always been with us.

With this in mind it is difficult to explain how Alzheimer's, a disease that now affects 20 per cent of people older than seventy-five, would not be noticed. As a contrast, much rarer conditions such as epilepsy – which affects only between 1 and 2 per cent of the general population (rising in old age and probably more common in AD) – have been noted for thousands of years.

"AMONG THE PEOPLE"

Few people older than the age of forty do not know what Alzheimer's disease is. Most can tell you of someone they know who has the disease or has died of it. In a national survey, 19 million Americans said they had a family member with Alzheimer's disease, and 37 million said they knew someone with the disease.

In North America and Europe, the numbers of people with Alzheimer's is staggering: one in ten persons older than sixty-five and nearly half of those older than eighty-five have the disease.[10] This translates to approximately 4.5 million Americans and 250,000 Canadians who have Alzheimer's disease today. In the United Kingdom, more than 700,000 people have dementia, about 55 per cent of whom have AD.

More significantly, the number of people with dementia is expected to increase steadily over the next twenty-five years.

- In Canada, 10,000 new cases of Alzheimer's are diagnosed each year – 27 cases per day.[11]
- By 2010, there will be about 840,000 people with dementia in the U.K. This is expected to rise to more than 1.5 million people with dementia by 2050.
- A study by Dr. Denis Evans of the Rush Institute on Healthy Aging and the Rush Alzheimer's Disease Center in Chicago shows that the prevalence of AD in the United States will increase 27 per cent by 2020, an astonishing 70 per cent by 2030, and nearly 300 per cent by 2050. His estimates indicate that by 2050, between 11 and 16 million people in the U.S. will have AD.
- Japan is one of the most rapidly aging nations in the world. The ratio of age sixty-five and older was 7.9 per cent in 1975 and 17.3 per cent in 2000. It is expected that in 2025 the rate will be 27.4 per cent, and the number of old people will be 33 million. As the population ages, it is also expected that the number of old people with dementia will increase – from 1.56 million in 2000 to 3.15 million in 2025, according to Japan's Ministry of Health, Labor and Welfare.
- Developing countries will see an even greater rise in numbers of people with dementia.

Nowadays doctors, scientists, and public-health officials routinely refer to the current "epidemic" of Alzheimer's that they assure us will get much worse in the coming years. Their choice of word is noteworthy: an *epidemic* almost always refers to a disease caused by some form of infectious agent – like the great flu epidemic of 1918 or the polio epidemics of the early 1950s.

Indeed, many scientists now believe that Alzheimer's too is due to an infectious agent. However, the suspected agent is not a virus or bacteria, but something far more deadly and sinister than either of these. In the next section of this book we will investigate a number of deadly diseases that may be closely related – and share the same cause – as Alzheimer's disease.

PART TWO

Kuru, CJD, Mad Cow, and vCJD

CHAPTER

KURU: THE CURSE OF CANNIBALISM

1953. Papua New Guinea. Nen lies by the fire on the mud floor of a primitive hut deep in the jungle. She has been dying in a particularly gruesome fashion for the last several weeks, and now family members gather around waiting for her final moment.

Her relatives are certain that she is the victim of sorcery. Perhaps Nen caused some real or imagined offence to one of the sorcerers of her tribe. If Nen knows of such an offence, she is beyond telling what it may have been.

Powerless against such magic, the victims, usually women and children of the tribe, pay a terrible price. Nen has suffered for seven months, and now she is dying of the shaking disease – kuru.

Nen's symptoms followed a course familiar to the rest of the tribe. First, she began to tremble, almost invisibly. Only the keenest observer would have noticed the slight shaking of her body, but Nen herself knew. Soon she was unable to disguise the symptoms. She could not walk without stumbling. She laughed uncontrollably but joylessly, or sometimes suddenly burst into tears. Her slurred speech sounded like the mumbling of a drunkard. After a few months her arms and legs often jerked spasmodically, sometimes as a result of a bright light or a sudden loud noise. She trembled continuously, and yet she was not feverish.

Before long she could not walk, could not feed herself, nor even talk. Eventually she could no longer swallow, and thus began the certain course of starvation.

As Nen lies dying, the other tribeswomen begin to prepare for her funeral. Not only will the dying woman be the "guest of honour" at the feast, she will also *be* the feast. Nen knows that she will be eaten after her death. She has already promised the best parts of her body to certain family members – an honour for both Nen and those who will consume her.

Shortly after Nen's death, the tribeswomen remove her body to the garden and begin their task. With sharp bamboo knives they dismember the body, skilfully removing muscles, heart, and other organs. Using a stone axe, they break the skull and scoop out the brain with bare hands, depositing it into a bamboo cooking tube. Nen's closest female relatives, along with her young children, will eat her brain and spinal cord, while the other women of the tribe will consume her other body parts. A few of the men may get a bit of muscle meat if they're lucky.

Not only will the tribe consume every ounce of flesh, they will also suck the marrow from the bones, then throw the bones in the fire. After a week or two of heat, the bones will be brittle enough to pound into powder, which the women will sprinkle over vegetables. Nen will be completely consumed by her tribe.

Such was the custom of the Fore (FOR-ay) tribe, at that time probably the only group of people on the planet who had institutionalized cannibalism as part of the rites of their society. This ritual display of mourning and respect for the dead, demonstrated by consuming their bodies, had been played out amongst the Fore tribeswomen for little more than half a century.

THE SHAKING DISEASE

The West Pacific island of Papua New Guinea, six times the size of England, is the world's second largest island after Greenland. It boasts the highest mountains in the world after the Himalayas. The snow-covered peaks slope down to steaming tropical rainforests, and the valleys between the mountains teem with huge rushing rivers. Although it has one of the longest coastlines in the world, it is almost impossible to land any-sized boat due to the high cliffs, dense mangrove swamps, and raging rivers.

Until the late twentieth century, one could find far better maps of the moon than of Papua New Guinea. In this huge, impenetrable, and inhospitable country lived hundreds, perhaps thousands of isolated tribes. As well as being isolated from the outside world, these tribes, because of the extremes of geography, were isolated from each other. Anthropologists, when they began to arrive in the early twentieth century, discovered an amazing diversity of languages and cultures. In fact, the natives of Papua New Guinea spoke a greater number of languages – 740 languages and 2,500 dialects – than the inhabitants of Europe.

It was to this inhospitable region that a young American doctor named Carleton Gajdusek came in 1957. He had just finished a two-year fellowship in Australia and before heading home to the United States decided to go looking for some adventure while studying primitive cultures – and where better to look than Papua New Guinea?

Trekking through the remote highlands, he met Dr. Vincent Zigas, a local medical officer who introduced him to the Fore tribe. The Fore occupied an area of about two hundred and fifty square miles. In this relatively small area (about the size of Edmonton, Alberta) lived about twenty thousand people.

At least two hundred people, or 1 per cent of the Fore population, were dying of a disease unknown elsewhere in the world, a disease that seemed to have started suddenly in the last one hundred years. So rampant was the disease, it accounted for more than half the deaths in the region. The natives called this gruesome disease kuru.

The symptoms started with a shaking and stumbling gait, rapidly progressing to slurred words, uncontrollable mirthless laughing, inability to walk or feed oneself, then to dementia, paralysis, coma, and death. From the onset of the first symptoms to death, the grim course of kuru took about six to twelve months.

Although the Fore were just one of hundreds of tribes living in the mountainous jungles of Papua New Guinea, they were the only people afflicted with this disease. Because it was restricted to this one tribe, it was first believed that kuru could be due to a genetic disorder, but fifteen times more women than men got the disease. Of the males who succumbed, most were children or young adults.

The Fore themselves believed that sorcerers casting curses caused the disease. Gajdusek suspected otherwise. The search for the cause of this disease uncovered one unique feature of the Fore – they were cannibals.

CANNIBALISM: THE LAST TABOO

Although cannibalism has been widely reported since prehistoric times, accounts of the practice have often been disbelieved, perhaps because of our inherent abhorrence of the idea.

Charges of cannibalism have often been used as a way of demeaning or dishonouring rivals and competitors. Early explorers in Africa heard widespread reports of cannibalism, but it was always someone else who practised it – our enemies, or the tribe just across the river. Hard evidence was almost impossible to come by.

In fact, the prohibition against cannibalism seems to be one of the few that transcends all races and all cultures. The practice is so abhorrent that legislators cannot imagine anyone would want to do it, and so cannibalism remains one of the only activities that, even though completely against societal norms and values, is not actually illegal in most countries.[1] Nevertheless, researchers have accrued enough evidence to confirm that cannibals have existed in most of the continents of the world.

As far back as the fifth century B.C., Greek historian Herodotus used the word *anthropophagy* (from *anthropos*, human being, and *phago*, eat) to describe the "savage customs" of the *androphagi*, man-eaters. Anthropologists still prefer this more descriptive term to cannibalism, which carries with it racist overtones.

Christopher Columbus encountered man-eaters in Guadeloupe among the Caribs, whose name in Spanish dialect was Canibales. Although Columbus made several errors interpreting the name Canibales (he associated it with the Great Khan of the Mongols), the name *cannibalism* has stuck with the practice ever since.

Anthropologists identify several types of cannibalism:

- *Auto-cannibalism* includes such benign habits as biting one's nails, licking the blood from a cut finger, or ingesting one's own mucus. More destructive auto-cannibalism practices include various pathological forms of self-mutilation such as eating pieces of one's own fingers or other body parts.

- *Ritual cannibalism* endures in the Holy Communion of certain Christian churches. Participants drink wine and eat bread or wafers, symbolizing the blood and body of Christ, as he urged at the Last Supper. Indeed, the widely held belief in transubstantiation – the conversion of the bread and wine into the actual body and blood of Christ – implies an acceptance of the cannibalistic practice. In the Anglican version of the Communion, the priest entreats the participants to drink the wine, and recites the words of Christ, "This is my blood, which is given for you."

- *Survival cannibalism*, eating human flesh in order to survive an ordeal, presents a unique dilemma to those who resort to it. When it's a case of do or die, people on the verge of starvation have often resorted to cannibalism.

  A Uruguayan rugby team, along with some members of their families, survived a plane crash in the Andes in 1972. Stranded in the mountains for seventy-two days, the sixteen survivors, in desperation, ate the bodies of their fellow passengers killed in the crash. Tellingly, they first ate the pilot, whom they did not know, before eating the bodies of their friends and relatives. Although the affair created a worldwide media sensation, most observers eventually came to the conclusion that the survivors had done the right thing in saving themselves.

- *Pathological cannibalism* refers to the eating of human flesh by psychotic individuals such as serial killer Jeffrey Dahmer or the fictitious evil doctor Hannibal Lecter from the film *The Silence of the Lambs*. Although he gained fame as a cannibal, Dahmer is known to have eaten only the bicep from just one of his victims.

- *Exo-cannibalism* is the practice of eating parts of vanquished enemies, a custom widely employed in the New World during the Aztec civilization of the sixteenth century. Aztec warriors killed and ate the brains and hearts of anywhere from 15,000 to 250,000 prisoners of war.[2]

- *Endo-cannibalism*, the practice of eating members of one's own tribe, flourished in the northwestern Amazon area, where Cubeo warriors' wives ate the penis of a dead victim to promote fertility. Some societies recycled the souls of their own villagers by drinking the ashes of naturally deceased relatives dissolved in corn beer or a banana drink.[3]

CANNIBALISM AND KURU

Endo-cannibalism as practised by the Fore may have had a practical function. Until about the mid-1800s, the tribe lived a tropical subsistence type of life; the women gathered beans, edible roots, fruits and berries, while the men mainly hunted wild pig, which was indigenous to the area. According to tribal custom, only men were allowed to eat pork and other meats. Women ate only fruits and vegetables.

Sometime in the mid- to late nineteenth century, women of the tribe started to prepare and eat the bodies of family members who had died. They discovered (apparently) that human meat tastes good and that it added much-needed protein to their diet. During the height of this practice in the 1940s, it was estimated that eating the 1 per cent of the population that died each year provided the nutritional equivalent of the pork in the men's diet.

After years of study under the most primitive conditions, which included performing autopsies by lantern light in crude huts while storms raged outside, Carleton Gajdusek concluded that kuru was the direct result of cannibalism. In fact, he showed that the disease could be passed to chimpanzees by injecting material from kuru victims into the brains of chimps.

The Fore abandoned the practice of cannibalism in the 1950s, and within thirty years kuru had all but disappeared. By then Gajdusek had formulated several criteria for kuru. The disease had five cardinal features:

1. It is transmitted by eating someone who had the disease.
2. The incubation period is extremely long. As many as thirty years elapsed between the cannibal feast and emergence of the disease. (Children and young adults who succumbed were more susceptible than older adults.)
3. While the disease is incubating, there are no warning signs that an individual is infected until the disease erupts.
4. Once the disease starts, there is a very rapid deterioration that always leads to death. No treatment is at all effective.
5. The victims' brains viewed under a microscope show a very distinctive pattern. Hundreds of tiny holes (voids) give the brain a spongelike appearance. Between the voids are areas of dense, amorphous material called plaques. Furthermore, some areas found in the cerebellum contain no cells, only brushlike plaques (also very dense), which are called spikeballs.[4]

Because the brains look somewhat like sponges, Gajdusek called this type of disease a transmissible spongiform encephalopathy (TSE). Gajdusek felt that there had to be some form of virus responsible for this condition, one which infected people very slowly. No known virus could do this, so Gajdusek proposed a new culprit, which he called a "slow virus." This would be a regular virus that for some reason took years and years before it began its attack on the brain. Such a virus had never been described before, and despite years of meticulous research to try to isolate the virus, none was ever found.

Gajdusek sent pickled brains of kuru victims to the United States for examination, where neuropathologist Igor Klatzo examined them. Under the microscope, Klatzo noted the knots of protein called amyloid plaques along with other brain damage – and became the first to make the connection between kuru and Creutzfeldt-Jakob disease (CJD). On September 13, 1957, Klatzo perceptively wrote to Gajdusek:

> I am afraid I am unable to give you any useful leads as far as etiology [cause] of this disease is concerned. It seems to be definitely a new condition without anything similar described in the literature. The closest condition I can think of is that described by Jakob and Creutzfeldt [CJD].[5]

There is no consensus about where the first case of kuru came from, but there is no doubt that all the subsequent cases were caused by the eating of infected flesh. Gajdusek's research proved that an infectious agent was involved and this agent could be transmitted orally.

For his groundbreaking work on this disease, Gajdusek received the Nobel Prize in Medicine or Physiology in 1976. He had so much to say about his fascinating work that his acceptance speech lasted two hours (compared to the usual forty-five minutes for most Nobel recipients). No one seemed to mind.

Twenty years after his Nobel win, Gajdusek's reputation suffered a serious blow when American authorities convicted him of pedophilia. Under a plea agreement, Gajdusek served one year in jail for sexual abuse.

Gajdusek continues his work on conformational changes in proteins. His groundbreaking scientific work on TSEs, however, would have far-reaching repercussions, and would ultimately lead to a Nobel Prize for another scientist.

CHAPTER 7

CREUTZFELDT-JAKOB DISEASE

Two doctors working independently in Germany first described Creutzfeldt-Jakob disease (CJD) in 1920–21. At that time, it was known as "spastic pseudosclerosis" or "subacute spongiform encephalopathy." A German doctor may have described it earlier in 1911, but the names of two neuropathologists, Dr. Hans Creutzfeldt (an assistant to Alois Alzheimer) and Dr. Alfons Jakob, have been attached to this disease since the 1920s. (It was first called Jakob-Creutzfeldt disease, and that name is still used in many major medical dictionaries. The disease described by Creutzfeldt in 1920 does not fit present criteria for transmissible spongiform encephalopathies [TSEs]. Consequently many believe that the honour of having described the first cases of TSE in humans should belong to Alfons Maria Jakob alone.)

All of the early CJD cases diagnosed had several common symptoms. Patients first exhibited psychiatric symptoms such as paranoia or emotionally inappropriate behaviour, loss of concentration, or poor memory. But no matter how the symptoms began, they always ended with dementia.

CJD: A VERY PECULIAR DISEASE

Imagine yourself a medical student in the 1960s, poring over your textbooks, studying the chapters on diseases. You would find diseases sorted into three groups:

- disorders you were born with, like muscular dystrophy;
- maladies you got as you got older, like heart disease; and
- diseases you caught, like measles.

By the time you finished reading all those chapters, you may have laid claim to a permanent seat in the library, acquired a lifelong addiction to coffee, and ruined your social life, but you knew most of what you needed to know about disease. Those chapters represented a vast store of medical knowledge that you would need in order to diagnose and treat any of the thousands of maladies that afflict human beings.

But at the very back of those 1,500-page textbooks you would find a final chapter called "Other Diseases." This was the chapter that featured diseases you might never see in a lifetime of clinical practice. This was the chapter that you and your fellow medical students would read if only to establish the claim that you had read everything.

In that final chapter, you would have found Creutzfeldt-Jakob disease. In 1957, when Carleton Gajdusek sent kuru brain samples to Igor Klatzo, CJD was so rare that Klatzo knew of only about twenty reported cases (which makes his observation of the similarities with kuru all the more remarkable). CJD remained at the back of the text for many more years – until the discovery of prions elevated CJD to a position among the most highly scrutinized diseases.

A Devastating Affliction

Like kuru, CJD produces severe physical damage to the brain – damage that leads to horrible physical and mental changes in the victim.

As devastating as the disease is for the victims, it's equally distressing for their families. They witness behaviour completely foreign to their loved ones. "He started running into the walls." "He became depressed and suicidal. His speech was slurring badly." "He became aggressive toward me." "He had a lot of difficulty lining food up with his mouth."

The victims confess to hallucinations, often reluctantly as they feel themselves sliding into a mental and physical abyss. Those close to them watch helplessly. "He saw trains. He thought we were riding in an airplane. He saw children singing in a choir, a lady in a rocking chair, a creek running beside his bed." Another case: "She said she saw a man jumping out of a window in the next building and started yelling and acting out. This was such uncharacteristic behaviour from a woman who never raised her voice to anyone."

Common to most cases are severe spasms and jerks of the limbs. "He told us he felt like he was falling, even though he was flat on his back in bed." "His arms were tied to the sides of the bed as they had been so bruised from a week of uncontrollable spasms." (In medical parlance, this is called *myoclonus*. *Myo* meaning muscle, *clonus* meaning jerk, describes a brief, sudden, singular, shocklike muscle contraction.) "She would be sitting up in the bed and her arm would just start to go up in the air and she wasn't even aware of it." "His hands were clenched across his chest and the jerks had caused him to scratch himself badly."

Sometimes the dementia doesn't come until near the end: "His spirits were high through all of this and he could still laugh at himself and make others laugh too." Another case: "Two weeks before he passed away, he knew who he was, who we were, what was going on, and what was being done."

But dementia seems inevitable. "After being admitted to the same hospital that he had gone to for many years, he became very confused. He thought I had taken him to a strange hospital in a town which had no hospital." "Less than a week later, he failed to recognize me."

"She started attacking people. We were having dinner one night and her neighbour came over and I went downstairs to take out the garbage and when I came up my son and her neighbour were in shock because she had attacked her neighbour and was yelling and shouting. They couldn't believe it. My son adored her, and he looked so scared. His sweet wonderful grandmother – where had she gone to? I couldn't answer that question for him because I didn't have the answer."

There is no cure for CJD. If there is any consolation for victims or their families, it lies in the fact that CJD usually lasts less than half a year.

Before long, sufferers fall into a coma, and die soon thereafter. "Suddenly he laughed out loud. He sat straight up in the bed with the most beautiful smile on his face and took a deep breath and said, 'Hell'! Then he laid back down and shortly thereafter he slipped into a coma."[1]

Because CJD occurred so rarely, it was sometimes misidentified. Victims and even their doctors often didn't know what had hit them. Diagnoses ranged from brain tumour, Lou Gehrig's disease (amyotrophic lateral sclerosis, or ALS), cancer, multiple sclerosis, stroke, arthritis, and inner-ear infections, to problems with eyeglasses, underactive thyroid gland, and vitamin B12 deficiency.

As victims deteriorate, their frantic families beg their doctors to do something. With little to guide them, doctors have tried Zoloft (antidepressant), Dilantin (to control seizures), Prednisone and Antivert (for vertigo and inner-ear infection), Aricept (a drug that has shown some promise in some Alzheimer's patients), Risperdal (a medication for schizophrenia), and B12 shots and pills. But nothing halts the relentless advance of CJD.

Until early in the twentieth century, the only victims exhibiting dementia were those who had late-stage syphilis or those with long-standing schizophrenia. A patient who became demented – and who had never had either of these conditions – attracted a great deal of medical attention. No one had seen this type of dementia that seemed to arise out of nothing. (A similar situation occurred in the 1970s when several young men with symptoms of multiple tumours of the lymph nodes or skin were diagnosed with Kaposi's sarcoma. Because Kaposi's sarcoma previously had been found only in elderly men, this occurrence attracted attention. It was this observation that led to the discovery of AIDS as a new disease entity.)

Early descriptions of CJD cases always included disorders of movement, either involuntary movements, tics, jerks, or twitches, or the opposite – rigidity of the joints or the inability to coordinate movement smoothly, resulting in a bizarre gait and eventually the inability to walk.

CJD usually appears in later life and runs a rapid course. About 90 per cent of patients die within one year. In the early stages of disease, patients may have failing memory, behavioural changes, lack of coordination, and visual disturbances. As the illness progresses, mental deterioration becomes pronounced and involuntary movements, blindness, weakness of extremities, and coma may occur.[2]

After analyzing more than one hundred case reports of CJD, Drs. Françoise Cathala and Henry Baron compiled a description of a typical CJD case.[3] It occurs in a man or woman between the ages of forty-five and seventy-five, most often in the late fifties to early sixties. About one-third of patients experience a prodrome, or early warning, consisting of overwhelming fatigue, change in sleep habits, and weight loss. These symptoms may be associated with headaches and vague, ill-defined pains. The onset of the disease is associated with mental deterioration in two-thirds of patients, and in one-half this may be the only sign. This mental decline may take several forms such as increasing forgetfulness, or difficulties doing what was once a routine job. Garbled speech and bizarre behaviour can occur. Sometimes emotions are affected, giving rise to terrible depression, apathy, or anger.

Two-thirds of patients exhibit some neurological sign such as involuntary movements or visual problems. However, all patients become demented. It was this symptom of dementia that first attracted all the attention, since dementia presented itself rarely in the general population.

We now recognize that victims may contract CJD in one of four ways. The route of transmission gives each form its name: sporadic, genetic, iatrogenic, and variant. We'll explore the first three of these forms in the balance of this chapter. Variant CJD is the subject of the chapter that follows.

Sporadic (or Classical) CJD

This is the most common form of CJD, representing about 80 to 85 per cent of the cases. It has been found in every country in the world where it has been looked for. It affects about one person per million of the population – the same rate of incidence as sporadic TSEs in other animals (beef, deer, elk, mink, and others).

Sporadic CJD tends to affect middle-aged and elderly individuals (average age around sixty-five), although this is not an absolute. Symptoms begin with a distinctly neurological illness that follows a very rapidly worsening course. The duration of sporadic CJD from onset to death is usually a few months (average four and a half months), and, in a few cases, just weeks, although this can also vary, occasionally up to a few years.

In the United Kingdom, fifty to sixty people die each year due to sporadic CJD. In the United States, there are about two hundred cases per year. Canada and Australia show similar rates. However, the officially stated mortality rate of about one person per million per year may be an understatement since CJD often escapes detection. In a Yale University study, 13 per cent of clinically diagnosed Alzheimer's patients were found on autopsy to have CJD.[4]

The cause of sporadic CJD remains uncertain. However, many individuals and scientific researchers now suggest that numerous cases of CJD may have been misdiagnosed as sporadic when in fact they resulted from specific causes. Several "cluster" cases of CJD have sparked further research and inquiry into other possible causes, including a genetic susceptibility. Some cases (perhaps many) were probably caused by eating of contaminated elk, deer, or squirrel, or by treatment with infected surgical instruments (which had previously been used unknowingly on CJD victims). Because these causes have not been well known previously, the origin has been listed as "sporadic." Many federal and local authorities are reluctant to admit the link between eating contaminated meat and contracting CJD. But a number of CJD victims are known to have consumed wild meat, like Doug McEwan, a Utah deer hunter, who died of CJD in 1999 at age thirty. Reports after his death noted that he handled deer carcasses often and regularly ate deer, elk, and antelope. The likelihood of dying of *sporadic* CJD at such a young age is extremely small.

Genetic CJD

In this very rare form (5 to 10 per cent of cases), an inherited abnormal gene causes CJD. In most cases of genetic CJD, family history reveals the presence of the illness. Some populations reveal a clustering of familial CJD cases. For example, Libya- or Slovakia-born Israelis show a 60- to 100-fold increase in incidence, an increase attributable to gene mutations rather than to transmission or environmental risk factors. Occasionally, genetic cases are seen in which no previous family history is identified. Blood tests provide the only definitive proof of the genetic abnormality. Genetic CJD may be as rare as one case in 20 million people.

Iatrogenic CJD

Diseases caused by medical intervention are said to be iatrogenic (from Greek, *iatros*, meaning physician). Iatrogenic cases represent less than 1 per cent of all CJD cases. Approximately four hundred such cases are known to have occurred around the world. Patients can acquire CJD iatrogenically from tissues transplanted from an individual infected with CJD. The most common infectious transplants include pieces of dura (the tough outer coating of the brain) that were taken from cadavers (thus called cadaveric dura mater) for use in some neurosurgical operations, and corneal transplants. More worryingly, the disease can also be transmitted by neurosurgical instruments that have been used in operations on patients with CJD. Even though the instruments have been sterilized by conventional methods, including autoclave, ethanol, and formaldehyde, they remain infectious. Finally, famously – or infamously – iatrogenic CJD has resulted from a treatment that seemed at the time to be almost miraculously beneficial to a small group of patients.

THE HUMAN GROWTH HORMONE DEBACLE

1958. Boston. Dr. Maurice Raben, an endocrinologist at the Tufts University School of Medicine, measures the height of his patient, a seventeen-year-old boy who had been diagnosed with hypopituitary dwarfism. During the past ten months, Raben had experimentally injected his patient with human growth hormone (hGH) obtained from human pituitary glands.

The human pituitary gland is an oval-shaped endocrine gland, about the size of a garden pea, situated at the base of the brain. The gland is divided into the back (posterior) and front (anterior) pituitary, each responsible for the production of its own unique hormones. The pituitary is sometimes called the master gland of the body because it controls the function of other glands in addition to secreting its own complex collection of hormones – hormones that govern growth, maturation, reproduction, and metabolism.

The pituitary gland secretes hGH, a microscopic protein, in short pulses with multiple peaks during the day, but mostly at night during the first hours of sleep, and after exercise. Although our bodies make hGH throughout our lifetime, it is more plentiful during youth. It helps to stimulate bone and muscle growth, which produces growth in children. It also plays an important role in adult metabolism.

When the pituitary fails to function correctly, an imbalance in hormones can create a number of problems. An excess of growth hormone will produce acromegaly, or giantism, as in the case of André "the Giant" Roussimoff, the famous wrestler from Grenoble, France. By the time he died at age forty-six, he had grown to 7 feet, 4 inches and 520 pounds.

Growth hormone deficiency (GHD) creates the opposite effect, and results in growth retardation characterized by short stature: arms, legs, trunk, and head typically are the same size in relation to each other as those of an average-size person (proportionate dwarfism). (Achondroplasia, the most common form of dwarfism, is a genetic condition that results in disproportionately short arms and legs compared to those of an average person. The average height of adults with achondroplasia is four feet.) Growth hormone deficiency may occur on its own (isolated GHD), or it may occur in conjunction with other pituitary deficiencies. Most cases of isolated GHD arise spontaneously or from an obscure or unknown cause, although they can also arise as a result of brain tumours or genetic causes, or from certain syndromes. The average adult height for untreated patients with severe isolated GHD is 56.3 inches (143 centimetres) in men and 51.2 inches (130 cm) in women.[5]

Raben's young patient was just over four feet (128 centimetres) tall when he began receiving injections of hGH. To the delight of the boy, his parents, and Raben, when they checked the measurement that day in 1958, they found that the boy had grown more than two inches (5.1 centimetres) in ten months. Further hGH injections over the next seven years would bring his total height to five feet, four inches (163 centimetres) – a leap of almost 14 inches (35 centimetres) from his height at the beginning of treatment.

Raben had accomplished a feat akin to a miracle – a triumph of medical research that would give new hope to thousands of abnormally short children and their parents.

When parents and pediatricians heard news of Raben's success they began to clamour for treatment. Because hGH was not available commercially, parents began enrolling their children in clinical trials. They soon discovered, however, that human pituitary glands, the *only* source of hGH, were not widely available. They began to petition hospitals and morgues to save pituitary glands from corpses. Pediatricians and endocrinologists joined the crusade.

In 1963, the National Institutes of Health (NIH) and the College of American Pathologists formed the National Pituitary Agency (NPA) in the United States. Canada developed a similar program through the Medical Research Council of Canada. Australia instituted the Australian Human Pituitary Hormone Program (AHPHP).

In Britain, the Medical Research Council ran the program as a clinical trial until 1977, when the Department of Health assumed full control. France and other countries in Europe and elsewhere set up their own programs. All of these groups supervised the collection of human pituitary glands and oversaw the distribution of the end product – human growth hormone.

Pathologists and mortuary attendants began harvesting pituitary glands from cadavers on a regular basis. Because the glands are so small, morgues commonly kept a bucket for collection. When the bucket was full – containing perhaps several hundred glands – it would be sent to the national agency in that country for distribution to labs. The labs processed the glands in batches, and extracted the valuable human growth hormone, which was then distributed to pediatricians and clinics.

For more than twenty years, the NPA supervised most of the hGH treatment in the United States. Because hGH was so scarce at first, patients taking part in clinical studies could receive treatment only for a limited time. Once they grew to a certain height, they had to give up the treatment and step aside to let others receive it.

Not until 1976 did commercial sources of hGH enter the market. By then there were at least a dozen methods of extracting and purifying the hormone. In 1977, the NIH awarded endocrinologist Albert Parlow, a research professor of obstetrics and gynecology at the University of California, Los Angeles School of Medicine, exclusive U.S. rights to process hGH with an improved Swedish-style filtration process. At peak production, Parlow would process as many as sixty thousand pituitary glands a year.

Each batch of hGH was derived from a pool of about ten thousand to sixteen thousand cadaver pituitary glands.[6] Each patient received hormones from two or three batches per year. By 1984, patients could receive treatment year-round and were allowed to continue to grow to whatever height they, their parents, and their pediatricians deemed appropriate.

And grow they did. Although hGH was still technically an experimental treatment, the NIH supplied it free of charge to pediatricians across America. More than eight thousand children in the United States and an estimated twenty-seven thousand worldwide received hGH treatment between 1958 and 1985. Some – perhaps many – of them experienced problems, but for the most part they grew and thrived, married, had children, and led normal lives that might never have been possible had they been left untreated.

But the wondrous little gland had other uses too. Infertile women all over the world could now receive treatment with human pituitary gonadotrophin (hPG), follicle stimulating hormone (FSH), and luteinizing hormone (LH). Many who were treated gave birth to multiple babies – twins, triplets, quintuplets. One Australian woman gave birth to nine, two stillborn and seven live, although sadly all seven died shortly after birth. Unknown numbers of persons may also have used hGH nontherapeutically; for example, athletes began to use it to enhance performance during intense physical training.

In the last year of distribution, about three thousand patients were being treated with hGH in the United States; slightly more than three hundred patients were being treated in Canada.

And then it all came crashing down.

On June 21, 1985, three unusual cases of Fatal Degenerative Neurologic Disease were reported in the *Morbidity and Mortality Weekly Report* issued by the U.S. Centers for Disease Control and Prevention (CDC).

Case 1. A twenty-year-old man began to have difficulty speaking, and developed a stumbling walk in May 1984. By September, his condition

had deteriorated so that he was no longer able to walk, could not care for himself, and required bladder catheterization. His mental abilities had deteriorated, and he could not carry on a meaningful conversation. He died in November 1984. An examination of his brain revealed holes in the brain cells consistent with Creutzfeldt-Jakob disease.

What could have caused this highly unusual condition? This patient had grown poorly during the first year of life. Hypothyroidism had been diagnosed when he was fifteen months old. In September 1966, doctors diagnosed growth hormone deficiency. The patient had been treated with daily injections of hGH from September 1966 to July 1980.

Case 2. A twenty-two-year-old man developed weakness and gait disturbance (stumbling walk) in the fall of 1983. During the next six months, his muscular coordination deteriorated, causing irregularity of muscular action in his arms, legs, trunk, and head. He also had speech impairment, difficulty swallowing, and dementia. He died in April 1985. A microscopic examination of the brain at the Armed Forces Institute of Pathology revealed extensive spongelike holes in the brain, similar to what is seen in Creutzfeldt-Jakob disease.

This patient had been evaluated for growth failure at seven years of age and was found to be growth hormone deficient. He had been treated with hGH from June 1969 through October 1977.

Case 3. A thirty-four-year-old man with hypopituitarism developed a stumbling walk in December 1983. During examination in June 1984, he showed an involuntary, rapid, rhythmic movement of the eyeball called bilateral horizontal end gaze nystagmus. He also exhibited a mild intention tremor (a tremor that arises or is intensified when a voluntary, coordinated movement is attempted) and wide-based gait. The symptoms worsened over the next several months, with increasing drowsiness, memory loss, and urinary incontinence.

The patient's symptoms progressed to include swallowing difficulties, double vision (diplopia), and, finally, dementia. He died in February 1985. No autopsy was done. He had received hGH from 1963 to 1969.[7]

Prior to these three cases, CJD was almost unheard of in people under the age of sixty. Like Alzheimer's disease, CJD was presumed to be a disease of the elderly. News of the three young victims, all of whom had received hGH, set off alarm bells everywhere.

When the first deaths were reported, the programs for collecting human pituitary glands and the clinical trials and treatment with hGH came to a screeching halt. After an urgent meeting on April 20, 1985, the NIH suspended the National Pituitary Program. Within forty-eight hours, Canada ended its program too. By June, Australia, Belgium, Britain, Finland, Greece, the Netherlands, and Sweden had joined the embargo.

Production of a new synthetic growth hormone, formulated in a lab rather than extracted from human pituitaries, became a top priority. By January 1986, all children with growth hormone deficiency in Canada were receiving biosynthetic growth hormone on a compassionate basis, and the children with newly diagnosed growth hormone deficiency received the same product over the next three years as part of a clinical safety trial.

Growth hormones had proved invaluable to patients, parents, and pediatricians, and the synthetic version quickly became the new standard worldwide. Today, synthetic hGH is used to reverse muscle-wasting in AIDS patients. It has become, in addition, one of the main drugs of abuse amongst athletes wishing to increase muscle size and strength. It is also promoted as an anti-aging treatment, but many of the claims made for hGH have a dubious scientific basis.

What had gone so terribly wrong with the human pituitary programs to create the tragedy of so many premature and horrible deaths? Some countries established inquiries to look into the disaster, while others ducked for cover. Lawsuits sprang up everywhere.

The NPA had issued only perfunctory guidelines for collecting human pituitary glands. The criteria for glands to be excluded from collection originally did not mention glands taken from patients who had died of CJD – at that time a little-known disease. Pituitaries taken from those who had suffered known systemic infections were banned, but pituitaries taken from those who died of chronic neurological diseases and other illnesses were accepted.

In early 1978, the NPA took the lead from Australia and specified that glands taken from patients with "viral dementia" – one of the names used for CJD and kuru – should be excluded from collection. Australia's exclusion criteria changed several times between 1966 and 1985, but not until 1982 did the criteria specifically exclude glands taken from patients who had died with "presenile dementia (Creutzfeldt-Jakob disease)."

Australia's Allars Inquiry later determined that the pathologists and mortuary attendants who removed the glands had not been informed of the criteria. No one institution or individual was made responsible for ensuring that the exclusion criteria were known or followed. Most of the pathologists and mortuary attendants contacted by the inquiry were unaware of any written exclusion criteria.

Furthermore, a pathologist could check for CJD only at the time of removal of the gland. No means existed to ensure that glands that were supposed to be excluded would not be collected. This was particularly so when glands were removed by mortuary attendants without supervision, "as was usually the case."[8]

In Canada, the United States, Britain, Australia, and other countries, the government programs had encouraged pathologists and mortuary attendants to remove pituitary glands by paying anywhere from twenty cents to two dollars for each gland. When commercial companies got into the business, payments skyrocketed to as much as fifteen dollars.

In the U.S., because of the well-supported drive to supply as many pituitary glands as possible for the manufacture of hGH, and because no one was considering the possibility of contamination by CJD, at least 140 CJD-infected pituitary glands may have been processed and randomly distributed among many hGH lots. The National Institutes of Health has denied liability, and has insisted (with some justice) that the deaths were unforeseeable.

The material that was injected had been derived from the pituitary glands of people who had died in hospitals and had come to autopsy. These cases had also had an incubation period of years to decades. This fact supported the theory that CJD was caused by an infectious agent, and that the infectious agent could be transmitted from person to person. In addition to being spread by ingestion (eating), as was the case with the Fore and kuru, it was now shown that it could be spread by injection. These cases also helped elucidate two other important facts about the infectious agent that caused the disease.

First, one or more of the pituitary glands from which hormones had been extracted had come from a person who had died of CJD. These infected pituitary gland(s) were mixed with typically ten thousand to sixteen thousand other non-infected glands. This large batch of glands was then processed to extract the hormones. As a result of this batching of such a large number of glands, the amount of material in any single dose of hGH or FSH/LH that came from the one infected gland must have been extremely small. This proved that, whatever the infectious agent was, *an extremely tiny amount of it was sufficient to spread the disease.*

Second, when these glands were batched and processed, the material underwent chemical and physical purification and cleansing that would have removed or killed any known bacteria or virus. Yet the hormones remained capable of infecting humans. This meant either the disinfecting process was flawed or – the more worrying conclusion – *the infectious agent was neither bacteria nor virus.*

THE SHEEP CONNECTION

On October 5, 1976, nine years before the first unusual cases of CJD appeared, Alan Dickinson, a veterinary geneticist and microbiologist, placed a phone call to the British Medical Research Council (MRC).

Dickinson, founder of the Neuropathogenesis Unit at the University of Edinburgh in Scotland, had been doing research on scrapie, an obscure neurological disease of sheep. Scrapie-infected sheep rub their flanks relentlessly (hence the name) against fences or trees, seemingly to relieve an intense itch. They also tremble, stagger, fall down, go blind, and invariably die – much the same as victims of kuru or CJD. Having experimentally infected mice with scrapie over a period of many years, Dickinson had observed that the pituitary glands became both infected and infectious.

One night, Dickinson abruptly sat up in bed with a terrible thought: that human pituitary glands could be infected with CJD. The next day he called the MRC to warn them of the danger posed by its growth hormone program. He followed up this call with a letter recommending that pituitary glands should not be collected from those who had died with dementia. While some at the MRC shared his concern, they essentially ignored his advice *for the next three years.*[9]

The MRC took more than a year to write to Dr. Carleton Gajdusek at the NIH, seeking his opinion about Dickinson's warning. But Gajdusek was travelling outside the United States when the letter arrived.

A visiting Australian pathologist named Colin Masters answered on behalf of Gajdusek, on May 8, 1978. "It would be reasonable to expect that the pituitary gland and/or surrounding tissue taken from a case of CJD disease would be contaminated with the virus," he wrote.[10]

So both the British (MRC) and the Americans (NIH) knew of the danger of using glands from organ donors who had died of CJD. Neither of them acted.

It took until 1979 before British officials finally agreed to support testing of various methods of gland processing and purification. It turned out that the Swedish method of filtration appeared to provide protection from CJD, but the results of the tests were not published until 1985.

The good news was that some four thousand American children who were treated after 1977, the year Albert Parlow took over U.S. production of hGH and insisted on the Swedish filtration system, were probably safe. The bad news was that many thousands more were now known to be at risk. More than 8,157 patients in the United States received hGH prior to the introduction of the synthetic version. As of April 2003, twenty-six of them had developed CJD. New Zealand reported five CJD cases among 184 people who received pituitary growth hormone. All five cases appeared in forty-six people who received hGH made in the U.S. prior to 1977. Of the 800 Canadians who received hGH between 1965 and 1985, none has developed CJD.

France, England, Holland, and Australia made their own hormone preparations. France has reported 89 CJD cases among 1,700 growth hormone recipients. In France, people who received hGH in 1984 and 1985 appear to be at highest risk for CJD. England has had 38 cases of CJD among 1,848 who received hGH treatment. Two people developed CJD in Holland. There have also been single confirmed cases of CJD in hGH recipients in Brazil and Australia.[11] Four Australian women developed CJD after receiving other pituitary hormones as fertility treatments.

As of April 2003, 161 cases of CJD worldwide have been related to growth and gonadotropic hormones of human origin, with about 18 of them diagnosed in the past two years.[12] Because of the extremely long incubation period for CJD, we can expect to see more cases arising in the future.

CJD has also been spread through other medical procedures. As of May 2002, iatrogenic (caused by medical intervention) transmission of the CJD agent has been reported in more than 250 patients worldwide. These cases have been linked not only to the use of contaminated human growth hormone, but also to dura mater (brain covering) grafts, corneal grafts, and to tainted neurosurgical equipment.

CHAPTER

MAD COWS AND DEAD ENGLISHMEN

Just as the growth-hormone scare was settling down, strange things began to happen in Britain down on the farm. The story of how a few dead sheep led to one of the most sensational public-health disasters of all time involves cannibalism in the feedlot and bureaucratic bungling that persists to this day.

"STENT FARM SYNDROME"

December 1984. Peter Stent, the owner of Pitsham Farm in Sussex, England, notices that one of his cows is behaving strangely. She is skittish, her back is arched almost like a cat, her head shakes even when she is standing still, and she has lost weight.

December 22. David Bee, a veterinary surgeon, is called to examine Cow 133 on Peter Stent's farm. He has no idea what he is seeing, so he seeks assistance from veterinary officer J.M. Watkin-Jones, a staff member at the local veterinary college.

February 11, 1985. Cow 133 dies.

By the end of April, seven more cows have died. As this seems to be an as-yet-unknown and new disease, the two veterinarians call it "Stent Farm Syndrome."[1]

In an effort to get a diagnosis, they send various specimens from the dead cattle to the veterinarian laboratory for analysis, but no definitive diagnosis can be made. A live cow with the disease is sent to slaughter so that the entire animal can be analyzed. Incredibly, it is slaughtered by shooting it in the head – so the brain is unavailable for dissection. This bungling is to become the hallmark of the entire story.

A second cow is sent off to the lab and this one is properly euthanized. Veterinary pathologist Carol Richardson examines the brain. Her account to Watkin-Jones in October includes a laboratory report showing fungal toxin in cattle feed at Pitsham Farm, and concludes that this cow, and probably all the others, died from some type of bacterial toxin. No further action is taken.

In June 1986, a nyala, a large African antelope, is put down at a wildlife park in England when it begins to exhibit unusual neurological symptoms. Martin Jeffrey, who works at the Central Veterinary Laboratory (CVL), a branch of the Ministry of Agriculture, Fisheries, and Food (MAFF), receives samples of the animal's brain and spinal cord. The brain has holes in it much like the holes seen in brains of scrapie-infected sheep. Jeffrey concludes that the animal died of a spongiform encephalopathy. Following this finding he writes a scientific paper intended for publication in the *Veterinary Record*, Britain's major veterinary journal. The paper is titled "Scrapie-like Disorder in a Nyala."

As is customary for all scientific journals, the paper was sent to a group of scientists considered experts in the field to assess its suitability for publication. Because scrapie was a well-known neurological disorder in sheep, and because it had created tremendous hardship for the British sheep farmers, the CVL wished to avoid even the mention of scrapie when referring to the new cattle disease. It was not until the following year that the chief veterinary officer for Great Britain, William Rees, discussed Martin Jeffrey's proposed paper with the director of the CVL, Dr. William Watson, and agreed that Watson would look at the draft and see if they could "avoid comparing the condition directly with scrapie" until more was known about the cause of the disease. Jeffrey had become aware of an embargo on references to scrapie in the context of the cattle disease, but was not prepared to amend his paper to delete references to scrapie. He later told the BSE Inquiry, "It would have been negligent to try to publish that without a reference to scrapie."[2] (After Jeffrey agreed to change the title to "Spongiform Encephalopathy in a Nyala" – thus omitting mention of scrapie – the article finally appeared in print in September 1988.)

Meanwhile, new reports of cattle suffering from a mysterious, scrapie-like disease found their way to the CVL – where senior staff avoided coming to conclusions that could damage the agricultural industry. Raymond Bradley, head of the pathology department, sent an internal memo to Watson and the CVL's director of research, Dr. Brian Shreeve, on December 19, 1986.

> The reasons for the interest are that the lesions observed have similarities to spongiform encephalopathies of other species and in particular scrapie of sheep. . . . I would advise keeping an open mind about the aetiology [cause] until we have more information. . . . If the disease turned out to be bovine scrapie it would have severe repercussions to the export trade and possibly also for humans if, for example, it was discovered that humans with spongiform encephalopathies had close association with the cattle. It is for these reasons I have classified this document confidential. . . . At present I would recommend playing it low key because a simple explanation may be forthcoming as a result of current investigations which will allay fears. You may also find the information valuable for the defence of the CVL in a political sense.[3]

In May 1987, neuropathologist Dr. Gerald Wells of the CVL gave a presentation to a closed joint meeting of the Medical and Veterinary Research clubs, informing them of this new disease in cattle. Wells was familiar with Rule 13 of the Veterinary Research Club, that matters discussed in the meeting should not be discussed with other people, or published in any form without the knowledge and agreement of the speaker. However, both Wells and Watson referred to the meeting as a watershed in relation to the dissemination of information to a wider audience.[4]

The numbers of cattle exhibiting odd behaviour grew exponentially. William Rees prepared a submission to the MAFF ministers. Because a general election campaign was underway, however, the submission was not received until mid-June. By this time, another zoo had reported that another African antelope, this time a gemsbok, had died of a spongiform encephalopathy. And similar symptoms had been detected in seven different cattle herds. Rees, in his memorandum to the government, remained cautious and protective of the cattle industry:

> There is no evidence that the bovine disorder is transmissible to humans. In the absence of such evidence, . . . it does not seem appropriate to impose restrictions on affected farms or on the sale of produce from cattle in affected herds. Irresponsible or ill-informed publicity is likely to be unhelpful since it might lead to hysterical demands for immediate, draconian government measures and could lead other countries to reject U.K. export of live cattle and bovine embryos and semen.[5]

In hindsight, a certain amount of hysteria may have been appropriate, and draconian government measures might have saved the day.

By July 1987, too many people knew too much, and the information began to reach the public. First, an article appeared under "Miscellaneous Conditions" in the *Veterinary Record* – but the term BSE was not used. In July, Rees informed a meeting of the National Farmers Union and the British Friesian and Holstein Cattle Breeder Societies of the emerging outbreak, which he now identified as bovine spongiform encephalopathy (BSE). Shortly thereafter, the CVL established a national database to keep track of new and existing BSE cases. In November, *Big Farm Weekly* published an article on BSE called "Mystery Disease."

Soon no one could deny that when the cattle brains were examined, they were found to look remarkably like the brains of people who had died of kuru or sheep that had died of scrapie. But MAFF feared linking these diseases publicly. They sent briefing notes to embassies and reiterated the need to distance BSE from scrapie.

> . . . [T]he fact that it so far appears to be a uniquely British disorder could prejudice our cattle exports if it is publicised

in inaccurate or exaggerated terms. It would be particularly misleading if it were to be described as "scrapie in cattle". Scrapie is a disease of sheep, the existence of which in British flocks is an impediment to our export trade, but although it is also an encephalopathy there is no evidence that BSE is attributable to the same cause as scrapie and it is important to distinguish between the two conditions.[6]

MAD MICE

The Neuropathogenesis Unit (NPU) in Edinburgh had been established in 1981 to do research into transmissible spongiform encephalopathies (TSEs) such as scrapie. Because the NPU was the main neurological research unit in the country looking into this class of diseases, it would have made sense for the Central Veterinary Laboratory to engage them from the beginning. The BSE Inquiry later criticized the CVL for their failure to invite the involvement of the Edinburgh unit at an early stage – when confirmation of the nature of the disease could have shortened the timelines and lessened the damage.[7]

In late 1987, the Edinburgh lab injected BSE-infected material from cows' brains into the brains of mice. When the mice came down with BSE, the NPU finally confirmed that BSE – bovine spongiform encephalopathy – is a TSE – a *transmissible* spongiform encephalopathy, that is, a disease similar to scrapie and kuru.

On October 25, 1987, more than two years after the first case was detected, the *Sunday Telegraph* in London ran the first story to appear in a national newspaper. The newspaper reported that an "incurable disease is wiping out dairy cows." By the end of the 1987, the official number of cases stood at 370 suspected and 132 confirmed – compared to the nine confirmed cases just six months earlier.

The next spring, a special working group, headed by Sir Richard Southwood, convened to study the implications of BSE, both in relation to animal health and to determine if there were any potential human health hazards. In 1988, the Southwood Working Party, as it was known, issued a statement that there would be minimal risk to humans as all infected cattle were to be slaughtered. The Southwood Report predicted there would be a maximum of seventeen thousand to twenty thousand cases of the disease in total. At least two of the group, Sir Richard Southwood and Dr. Hilary Pickles, had reservations about whether the number of cases of BSE had plateaued – reservations that ultimately proved justified. The Working Party did not, however, feel that they were in a position to dissent from the conclusions drawn by the experts they consulted.[8]

At the same time, the government started to develop a policy to slaughter diseased animals and to compensate farmers for the slaughtered cattle. Prior to this time, diseased cattle were killed, but the farmer received no compensation – another bad decision on the part of the government. With no compensation, farmers often hastened to ship a cow off to market at the earliest sign of disease – to ensure at least some return on their investment.

CANNIBAL COWS

Cattle are herbivores. Left to their own devices they would never think of eating dead cows – or any other kind of meat. They prefer grass even to more dense nutrients such as grains.

In 1988, scientists at CVL decided that animal feed practices were to blame for BSE. Manufacturers of cattle feed (and most other animal feeds) added protein-rich material to grain-based feed with a view to increasing milk production and fattening the animals more efficiently. This added material, known as meat and bone meal (MBM), was derived from cattle and other animals that for various reasons were deemed unfit for human consumption. The major reason that an animal would be so designated was that it was sick or had died of some disease rather than being slaughtered in the normal manner. Logic suggests that if you feed parts of cattle that have died to living cattle, there is an excellent chance that whatever agent was responsible for killing the original cow will remain in the food chain and affect other cattle.

The potential link between MBM in feed and BSE led to a ban on feeding cattle with material derived from other ruminants (grass-eating animals, i.e., sheep and cows).

In 1989, the MAFF imposed a total ban on human consumption of "certain cattle offal" known officially as Specified Bovine Offal (SBO).

SBO consisted of brain, spinal cord, spleen, thymus gland, and tonsils. Shortly thereafter, pet-food manufacturers agreed to a voluntary ban on using the same material in their products. In July 1989, the European Union imposed a ban on the export of all British cattle born before July 18, 1988, and on offspring of affected or suspected animals.

The number of new cases was growing at an alarming rate. By the end of 1989, the United Kingdom recognized 7,228 confirmed cases of BSE. By the end of 1990, there were more than 14,000 new cases.

Government inspectors found that almost none of the meat-packing plants were in complete compliance with the SBO ban. Later it was found that cross-contamination occurred at renderers, in feed mills, and on farms, as small quantities of ruminant protein, some of it consisting of SBO, were being mixed into cattle feed. The later BSE Inquiry noted that the primary causes of this "sorry story" were "not so much short-comings in monitoring the Regulations, but a lack of foresight when the Regulations were introduced, coupled with false assumptions as to the size of a fatal dose."[9]

There was at that time a widespread belief that a cow would have to eat a substantial quantity of infective material before it could become infected with BSE. It was not until early 1995 that experiments demonstrated that a cow need eat only one gram of infective brain tissue to become infected. The volume of one gram is equivalent to two peppercorns.

The press began to report in detail the specifics of the outbreak. *The Independent* newspaper reported that some scientists working in NPU stated that there was "a remote possibility" that BSE could be contracted by humans. Another scientist said that nothing could induce him to eat sweetbreads (or thymus – an organ meat), spleen, or brain. The technical director of the British Meat and Livestock Commission said, "A human would have to eat an impossible amount of pure cow brain at the height of infection to reach an equivalent dose [needed to infect a cow], and even then there is no evidence that the disease would infect humans."

MAD MICE II

Early experiments with mice had shown they could contract BSE when researchers injected infected material directly into their brains. By

February 1990, scientists had succeeded in passing the disease to mice through infected food, thereby proving that species not even distantly related to cattle could get the disease. The MAFF called these "unnatural methods of infection, which can only be done experimentally in laboratory conditions and which would never happen in the field." They further declared that the results provided evidence that ". . . BSE behaves like scrapie, a disease which has been in the sheep population for over two centuries without any evidence whatsoever of being a risk to human health."

Non-governmental scientists, however, were saying something quite different. Dr. Helen Grant, a retired neuropathologist, appeared on television and predicted that humans could be hit within twenty years. Worse, she said that infected cattle offal was still being used in pies and meat products. "My gut feeling," Grant said, "is that some genetically susceptible people may have become infected with material by eating meat products." The press also quoted Professor Richard Lacey, a clinical microbiologist at Leeds University, and a consultant to the World Health Organization, as predicting that ". . . in years to come our hospitals will be filled with thousands of people going slowly and painfully mad before dying."[10]

The public did not know what to believe. Was this, as the government maintained, just another variant of scrapie, a disease confined only to one species, something that might cause problems for farmers, but posed no risk to humans? Or was it, as Lacey feared, a new and virulent disease that could cross species barriers, a disease that posed a real risk to humans? As the debate raged, several school districts decided to ban beef from their menu. MAFF and the Department of Health worried whether other bulk purchasers such as the armed forces, prisons, and hospitals might follow suit and ban British beef.

Then the cat died.

MAD CATS

The ban on SBO in pet food was still voluntary at this point. In May 1990, government ministers learned that veterinarians at the University of Bristol had diagnosed a "scrapielike" spongiform encephalopathy in a five-year-old male domestic cat. This type of disease had never before been diagnosed in cats.

Before the year was out, twelve cats had been diagnosed with what was now called feline spongiform encephalopathy (FSE). The new diagnosis came as a huge blow to a government already reeling from the BSE crisis. The CVL noted that ". . . once this information becomes public (as is inevitable), it is bound to excite comment and a connection will be made with BSE, scrapie, and possibly even with CJD." Indeed it did.

There was a huge amount of public fear and uncertainty. If the disease can occur in cats, why not people? The government tried to spin the news in a more favourable manner. First, they issued statements to the effect that this disease was probably always present in cats but rare, so no one noticed. The ministry insisted there was "no likely connection between this case and BSE." A mass-circulation newspaper, the *Sun*, published a story that said BSE could be the biggest threat to human health since the Black Death plague in the mid-fourteenth century. The government's reaction was again to state that there was no danger.

But Germany refused to import British beef for human consumption, declaring that it posed too great a risk to human health. Beef consumption in Britain reached its lowest levels in decades. Sixty-five per cent of doctors said they had changed their beef-eating habits due to fears of BSE. Clearly, the British beef industry was in crisis.

MAD HUMANS

Up to this time, there were only three known spongiform encephalopathies in humans – kuru, Creutzfeldt-Jakob disease (CJD), and the very rare Gerstmann-Sträussler-Scheinker disease (GSS), all of which caused neurological damage leading to motor dysfunction and dementia. On autopsy, all of these diseases showed a spongelike deterioration of the brain tissue.

The Southwood Working Party had earlier published their view that there was almost no risk that humans would contract the human equivalent of BSE by eating infected meat. But hedging their bets, they noted that humans were susceptible to some spongiform encephalopathies and added:

> It is likely that cattle will prove to be a "dead-end host" for the disease agent and most unlikely that BSE will have any

implications for human health. . . . [But] it is a reasonable assumption that were BSE to be transmitted to humans, the clinical disorder would closely resemble CJD. Depending upon the route of transmission, the incubation period could be as little as a year . . . or several decades.[11]

With this warning in mind, the government set up the CJD Surveillance Unit with the mandate to identify any change in the number of cases of CJD, and to determine if there was any connection between CJD and BSE. The CJD Surveillance Unit project began on May 1, 1990, at the Western General Hospital in Edinburgh.

In March 1993, Dr. Robert Will, then director of the CJD Surveillance Unit, published a report in *Lancet*, one of Britain's most prestigious medical journals, describing the case of a farmer who had died of CJD. One of the farmer's dairy herd had been infected with BSE. Will concluded: "CJD in our case is most likely to have been a chance finding, and a causal link with BSE is at most conjectural."[12] In August, the *Daily Mail* recorded the death from CJD of another dairy farmer, aged fifty-four, who had also had BSE in his herd. The Department of Health issued a statement:

> The Government's expert advisers have considered this case and have agreed that there are no features that give cause for undue concern. The symptoms of CJD in this case were entirely consistent with the development of the sporadic form of CJD, and there was no evidence that the patient had been exposed to animals associated with BSE, or to their products, any more than many other farm workers involved in animal husbandry. Since the illness of the cows and the patient occurred within months of each other, the animals and the patient had presumably incubated disease in parallel for some years. It is most unlikely therefore that there is any direct link between the cases of BSE and the occurrence of disease in the patient.[13]

In other words, these farmers were just people who were unlucky enough to be the one-in-a-million people who get the disease. There was no link between CJD in humans and BSE in cows.

In December, a third farmer died of CJD. And the news would get worse.

By 1996, the number of new BSE cases reported in cattle was dropping rapidly – more than eight thousand new cases compared with the high of more than thirty-seven thousand new cases in 1992. This should have been good news for the government, but as the number of cases of BSE in cattle was falling, the number of cases of CJD in humans, especially young people, was rising.

But the CJD seen in these young people exhibited a different pattern of damage than the usual CJD pattern. The unusual characteristics included a relatively prolonged duration of illness as compared to sporadic CJD, early psychiatric symptoms, prominent ataxia (loss of control of movement), and the absence of the characteristic EEG appearance seen in sporadic CJD. One of the neuropathological features common to several of the cases was large plaque deposits consisting of dense prion protein deposition in all areas in the brain. Dr. Robert Will considered that these cases were unusual, because only about 10 per cent of cases of sporadic CJD had plaque deposition. He wondered about the possibility of a genetic form of CJD. Furthermore, the usual sporadic form of CJD affected mainly older people – usually around sixty-six years of age. This CJD struck young people in their teens and twenties.

In February 1996, Will was still not persuaded that beef was unsafe.

> My view is that there is a remote theoretical risk that BSE in cattle might cause disease in humans. The risk from beef and beef products is likely to be negligible, provided statutory measures are fully enforced. I would also agree with the statement that there is currently no scientific evidence that BSE can be transmitted to humans or that eating beef causes CJD.[14]

The key to whether the seemingly new type of CJD was related to BSE would be determined by comparing cases of young deaths from CJD in other countries. Was the fact that the victims were younger than usual CJD victims the reason for the unique symptoms and pathology of the disease, or had they in fact died from a different kind of CJD than had been known before? If this were a new kind of CJD, not seen in young people elsewhere, odds increased greatly that BSE-infected meat, a uniquely British problem, would be the culprit.

Enough young people had died of CJD in continental Europe to make a comparison with the U.K. cases. After gathering data from other countries and searching literature worldwide, Will ascertained that none of the young cases in continental Europe or elsewhere had the unique clinical and pathological characteristics of the young cases in the U.K. In March, he concluded that the new type of CJD was restricted to the U.K., "consistent with a causal link with BSE."[15]

This new form of the disease became known as a new variant of CJD, or vCJD.

THE BEEF CONNECTION

The cases of Stephen Churchill, Alison Thorpe, and Vicky Rimmer were typical of the new form of CJD.

Stephen Churchill always got good grades. He was bright, popular, and good at sports. His sister Helen described him as "the life and soul of any party." When he was thirteen, he joined the local squadron of the Air Training Corps (ATC). At eighteen, he was doing his A levels in school so he could join the Royal Air Force.

That year, his parents noticed he became quiet and withdrawn. On a number of occasions, he seemed drunk and uncoordinated. His schoolwork deteriorated, and he started saying things that made no sense. Helen was dismayed in November 1994 when Stephen claimed he had a job at a jeweller in town, a jeweller that she knew did not exist. He also said he was going to work at a pub, but he was very confused about the times he was expected to work. His doctor prescribed an antidepressant drug.

Stephen began to lose weight, suffered from auditory and visual hallucinations, walked unsteadily, and became lethargic. Soon he could not dissociate between reality and fantasy, and became frightened by the actions of cartoon characters or actors on television. He became clumsy, and needed help to go to the bathroom and to dress himself. He also needed support when walking.

Stephen underwent a barrage of tests and investigations including an electroencephalogram (EEG), a lumbar puncture for analysis of his cerebrospinal fluid, a cranial CT scan, a cranial MRI scan, and various blood tests. All tests proved either to be normal or inconclusive.

Helen bought Stephen a pair of fluffy gorilla slippers for Christmas. To her horror, they frightened him so much that he went to bed for the remainder of the day. In January, he was admitted to a psychiatric hospital.

In March 1995, doctors told his parents that Stephen might have CJD. Just ten months passed from the time he became ill until Stephen became completely unable to walk or communicate and needed round-the-clock care. In May, two weeks after he went into a nursing home, he died. He was nineteen.

October 1996. Macclesfield, England. Alison Thorpe's husband, Richard, notices that Alison's mood has changed quite suddenly. She has no time for him, and has become very short-tempered. She suffers from migraines two to three times a month, each time staying home from work for two to three days.

On a family holiday in July 1997, both Richard and his brother-in-law Stephen notice that Alison has lost a lot of weight. In particular, Alison's legs have become very thin. Within a week of her return to work after her holiday, she complains of feeling very tired and says that her legs are aching.

October 1997. Alison cries a great deal, and shouts at people for no reason.

Christmas 1997. Alison has become very "wobbly" on her feet. She also begins to suffer from panic attacks, during which her whole body shakes incessantly.

January 1998. Alison stumbles frequently, and has difficulty going up and down stairs. Her physician diagnoses stress and prescribes Prozac.

February 1998. Alison develops a shuffling gait and holds on to furniture for support. She seems unable to stand up straight, and lists to the right.

April 1997. Alison begins to suffer from visual hallucinations – she sees snakes and spiders in her bed. She has difficulty swallowing food, and feels pressured to eat by medical personnel who believe she has anorexia nervosa.

May 1998. Not long after her twenty-fifth birthday, Alison is admitted to hospital under the care of a consultant neurologist. The neurologist suspects new variant Creutzfeldt-Jakob disease (vCJD) but tells Richard that he can't be 100 per cent certain until brain samples are microscopically examined following a post-mortem examination.

June 1998. Richard takes Alison home and alternates with family members giving her constant care. Over the summer, Alison deteriorates rapidly, both mentally and physically. She becomes doubly incontinent. Her speech is incoherent, and she is totally immobile. She perspires profusely, although she is cold to the touch.

August 27, 1998. Alison Jane Thorpe (née Hodge) dies at home. Three weeks later, the CJD Surveillance Unit in Edinburgh confirms that Alison died of the new variant form of CJD.

At age fifteen Victoria (Vicky) Rimmer was transformed from a perfect teenage daughter into a moody, depressed wreck. Her grandmother later told how she first noticed symptoms in her granddaughter. "She started losing weight and started to look anorexic. She started falling, like you see cows with BSE staggering on television. . . . She couldn't understand and said, 'What's happening to me, Mum?'" Admitted to hospital in August 1993, Vicky went blind, then fell into a coma and never regained consciousness. She died four years later in November 1997.

After Vicky became comatose but before she died, the chief medical officer for the Department of Health released a statement in which he declared, "On the basis of work done so far, there is no evidence whatever that BSE causes CJD and, similarly, not the slightest evidence that eating beef or hamburgers causes CJD."[16]

Professor James Ironside, from the CJD Surveillance Unit, later told the inquest into Victoria Rimmer's death that her brain was so damaged that at autopsy it looked like the brain "of a 90-year-old who had undergone severe neurological damage."

"MAD COW CAN KILL YOU"

The early research on kuru paved the way for scientists to make the con-nection between mad cow disease (BSE) and CJD. Kuru showed that can-nibalism was responsible for transmitting the infection. The advent of BSE in the late 1980s and early 1990s proved that infection could be transmitted by one species eating another. Eating the meat of (now-cannibal) cows infected with BSE proved to be the cause of the new vari-ant Creutzfeldt-Jakob disease (VCJD).

But VCJD does not behave like most other infectious diseases. Whereas the incubation period for most infectious diseases is measured in days or weeks, the CJD incubation period (the time between being exposed to the disease and demonstrating symptoms), like that of kuru, tends to be years, possibly from five to fifteen years in the case of chil-dren, and perhaps decades in the case of adults.

Britain did not identify the first cows known to have died of mad cow disease until 1985. Beef eaters may have been exposed to infected meat from several years before that date. In 1989, cattle brain, spinal cord, spleen, intestines, thymus, and tonsil were removed from food for human consumption in Britain. But not until 1996 did the doctors and scientists see enough cases of VCJD to make the connection between this new form of the rare disease and the eating of cows infected with BSE. In March that year, the British government reluctantly announced that there could be a link between CJD and mad cow disease. By this time, countless consumers may have been exposed to contaminated beef.

On March 20, 1996, the headline in the *Mirror* blared, "OFFICIAL: MAD COW CAN KILL YOU." On that day, the Spongiform Encephalopathy Advisory Committee (SEAC), which had taken over the role of the Southwood Working Party, issued a statement that became the basis of the government announcement. In separate speeches, the ministers of health and agriculture admitted they had identified ten cases of CJD in young people. In words that would shock the world, they stated, "On current data, and in the absence of any credible alternative, the most likely explanation at present is that these cases are linked to exposure to BSE before the introduction of the SBO [specified bovine offal] ban in

1989. CJD remains a rare disease and it is too early to predict how many further cases, if any, there will be of this new form."[17]

Many people had long believed that the risk of transmission of BSE to humans was much greater than the government's assessment, and did not hesitate to say so. For them, the announcement came as a vindication of their warnings, albeit an unhappy one.

Professor Richard Lacey and Dr. Stephen Dealler were two scientists who, for much of the time, worked together in considering and writing about the implications of BSE. They were particularly concerned about the food risks posed by meat from infected animals that were not yet presenting noticeable symptoms. Lacey had been a consultant to the World Health Organization, and an adviser to MAFF on preventing diseases in people that are derived from animals and food. Dealler is a consultant microbiologist and had previously been a medical senior registrar in Lacey's department.

Dealler recently noted that the United Kingdom does not expect to see the last case of BSE until after 2006. The rate of new cases of BSE in cattle peaked at around 37,000 in 1992, with a nearly similar number in 1993, when it began declining slowly for a few years, and then more quickly. At peak, BSE affected 0.3 per cent, or three cows in every thousand of the national herd. (In 2002, 755 cases were reported in the U.K., including 445 new cases for the year. At the time of writing, approximately 30 cases per week are identified, although this number continues to drop.[18] But even as the infection rate declines in the U.K., the disease is now rising rapidly on mainland Europe. Despite the fact that meat and bone meal (MBM) had been banned as cattle feed in the U.K., the U.K. exported 25,000 tons of it to Europe in 1989. Because of this indifference to the possible human-health consequences (it's no good for us, so let's ship it to someone else), the European BSE epidemics may not be over until after 2010.

As of February 2, 2004, the total number of people listed as definite and probable vCJD cases (dead and alive) in the U.K. stood at 146. The number exposed could be in the millions or hundreds of millions.[19] Other countries have reported much lower numbers: six in France, and one each in Canada, the United States, Ireland, and Italy. However, few countries have surveillance systems; therefore the geographical distribution of the incidence of vcjd needs to be better defined.

Stephen Dealler summed up the seriousness of the current situation in 2001:

> In the U.K., one of the major problems was that we did not know which cow was infected before it had any symptoms. As a result we ate six out of every seven of them. This represents over 800,000 infected cattle entering the human food chain and the population eating 50 meals each made of their tissue.
>
> So far, 106 cases of vcjd have been identified, with between 1,000 and 10,000,000 potential cases incubating the disease. We don't know who to treat, whose blood to avoid at blood transfusion, who represents a risk at surgery or dentistry, or which asymptomatic cattle to dispose of.[20]

Indeed, we do not, as we continue to discover.

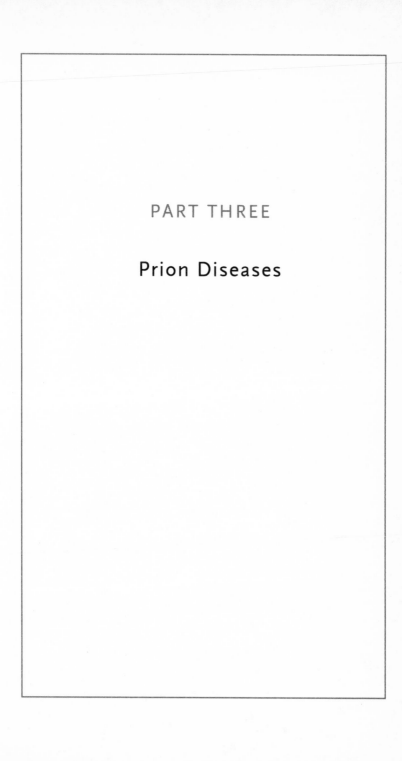

PART THREE

Prion Diseases

CHAPTER

MINUSCULE ASSASSINS

If a group of scientists were asked to design the ultimate biological weapon of mass destruction, what would they come up with? The perfect weapon might possess the following properties:

- A single exposure to the agent should be sufficient to cause the disease.
- The agent should be tough enough to withstand conventional disinfecting methods.
- It should lie dormant in its victim long enough that the entire population is infected before the first case develops symptoms, and while lying dormant should be undetectable.
- The agent, once introduced, would perpetuate itself in the food chain so that a small amount could infect the entire population.

In the 1980s, the British agricultural industry, with help from the British government and Mother Nature, inadvertently not only propagated such an agent but also let it loose upon the world.

A HUMOROUS NOTION

Prior to the fifteenth century, infectious diseases were viewed as evidence of some deity's displeasure. People who were affected offered sacrifices to attempt to appease the gods who had caused them to fall ill. Lo and behold, sometimes the sacrifices worked, and you got better – but sometimes they didn't work, and you died.

"Rational thinkers," beginning with the Greeks, dismissed the idea that some divine intervention either caused or cured disease. They formulated a theory that not only explained disease but health and human nature as well. This was the theory of the "humours." This precept held that four cardinal humours reside within every individual. (*Humour* is from the Latin word *umor*, meaning moisture or fluid.) The four humours were:

- blood;
- phlegm;
- choler (also called yellow or green bile); and
- melancholy (also called black choler or black bile).

Greek physician Galen (*circa* A.D. 130–201), physician to Roman emperor Marcus Aurelius, and heir to Hippocrates, taught the importance of maintaining balance between these four bodily fluids. Health consisted of having all four in the proper balance, while disease was the result of an imbalance – either too much or too little of one or more of the humours. Physicians used different means to restore harmony or correct the equilibrium in those four humours – emetics (herbs or medicine to make one vomit), cathartics or purgatives (laxatives), or bloodletting. Because many disorders were felt to be due to too much blood, bleeding was a mainstay of treatment for most conditions.

This physiological theory influenced European medicine until well into the nineteenth century. Some have suggested that the acceptance of the theory of humours retarded genuine physiological research since the humours could be said to account for almost any condition.

People's personalities could also be understood in terms of which humour predominated in that individual. Each fluid promoted a specific personality characteristic. Thus people with a preponderance of blood were sanguine, characterized by a ruddy complexion and a brave and amorous personality, while others might be phlegmatic (sluggish), choleric (quick to anger), or melancholic (depressed). It is testament to the enduring power of this system that we still use those four words today to describe various personality types.

LITTLE EELS

Not everyone bought into the system of humours. This moist concept of medicine began to dry up in the sixteenth and seventeenth centuries due to the work of three men: Robert Hooke (1635–1703) from the Isle of Wight in England, Antonie van Leeuwenhoek (1632–1723) from Delft in the Netherlands, and Louis Pasteur (1822–1895) from Dôle, France.

Hooke, a chemist and physicist, is often credited with the invention of the microscope. Historians suggest that he probably did not invent it, but he was the first to build several working instruments and to popularize their use. His book *Micrographia* (1665) was the first to describe it.

At almost the same time in Delft, a cloth merchant named van Leeuwenhoek refined his microscope. During his life he ground hundreds of lenses, making ever more powerful instruments, his crowning achievement being one with a linear magnifying power of 500 and a resolving power of one-millionth of a metre. He used his microscopes to examine and describe everything from cloth to insects. In September 1674, he examined some cloudy water taken from a local pond.

> I saw very plainly that there were these little eels, or worms, lying all huddled up together or wriggling: just as if you saw, with the naked eye, a whole tubful of very little eels and water, with the eels a-squirming among one another, and the whole water seemed to be alive with these multifarious animalcules. This was for me, among all the marvels that I have discovered in nature, the most marvellous of all; and I must say, for my part, that no more pleasant sight has ever yet come before my eye than these many thousands of living creatures, seen all alive in a little drop of water, moving

among one another, each several creature having its own
proper motion . . .[1]

He described this marvellous discovery in a letter to the Royal Society
(of Science in London) on October 9, 1678. The sheer smallness of these
creatures generated much skepticism. He wrote that there could be well
over thirty thousand creatures living in a drop of water no larger than a
mullet seed. He further observed that these tiny animals had all of the
bodily organs required for the life that it led. As these were complete,
formed living organisms they must reproduce after their own kind as
the Bible indicated, and not arise spontaneously from dirt or other
decaying substances as many people believed at that time.

Later van Leeuwenhoek undertook a number of experiments with
saliva and particles found between the teeth (both his own and others').

The number of these Animals in the scurf of a mans Teeth,
are so many that I believe they exceed the number of Men in
a kingdom. For upon the examination of a small parcel of it,
no thicker then a Horse-hair, I found too many living
Animals therein, that I guess there might have been 1000 in a
quantity of matter no bigger then the 1/100 part of a sand.[2]

Van Leeuwenhoek had discovered the world of bacteria. He had
opened the doors to new worlds of microbiology, embryology, histology,
entomology, botany, and crystallography. What role these bacteria played
in human affairs, especially disease, was not appreciated for another
hundred years, until the work of the first microbiologist, Louis Pasteur.

Despite van Leeuwenhoek's work, the belief in spontaneous generation as expounded by Aristotle still persisted. It was common belief that
flies, maggots, worms, and other small creatures arose spontaneously
from putrefying matter. One could easily prove this theory by leaving a
piece of meat out on a table; within a few weeks all forms of small creatures would appear on it.

This theory offended Pasteur's religious beliefs, as it seemed to
remove the necessity of God from the creation of life. In a classical
experiment, Pasteur took two flasks filled with a broth. He heated both

flasks to the boiling point, then sealed one so that it was airtight and left the other open to the air. The contents of the open flask underwent fermentation while the sealed flask did not. He thus demonstrated that airborne yeasts, not spontaneous generation, caused the fermentation.

By the mid-1800s, Pasteur was the most famous scientist in France, so when a mysterious disease that was killing silkworms threatened the French silk industry, Pasteur was commissioned by the government to investigate. He discovered that a tiny parasite caused the disease, and recommended that all infected worms be destroyed at the first sign of the disease. This very simple remedy eradicated the problem and saved an important industry.

His work with the silkworms piqued his interest in the entire subject of infectious diseases, leading to one of the greatest medical insights of all time, the germ theory of disease. This theory stated that germs (bacteria) cause infectious diseases, and that different germs cause different diseases. This ideology led to two of the most important public-health measures of all time: sterilization and pasteurization. These two measures have saved countless lives in the ensuing century and a half.

INFECTIOUS PARTICLES

Building on Pasteur's germ theory of (infectious) diseases, the great German bacteriologist Robert Koch (1843–1910) formulated a universal method by which to prove experimentally whether a disease is infectious. This scientific method consists of a set of precepts known as Koch's postulates, and these remain the basis of all experimental work in infectious diseases.

- The causative agent (germ) must be present in every case of the disease and not be present in healthy animals.
- The pathogen (germ) must be isolated from the diseased host animal and must be capable of being purified so only a single pathogen is present.
- The same disease must be produced when a healthy animal is exposed to the purified pathogen.

- The same pathogen must be recoverable from the animal
  that was artificially infected, and this pathogen must be
  capable of being purified.

Using these guidelines, scientists now had a valuable and universally acceptable method of proving whether a disease was infectious, and also of figuring out what the infectious organism (germ) was.

Van Leeuwenhoek had noted that the bacteria he observed under his microscope were alive. They moved, they fed, they excreted waste products, and they reproduced.

The great Irish satirist and poet Jonathan Swift (1667–1745) summed up the general view of the germ theory:

> So, the naturalists observe, the flea,
> Hath smaller fleas that on him prey;
> And these have smaller still to bite 'em;
> And so proceed, ad infinitum.

British mathematician Augustus De Morgan (1806–1871) later tried his hand at improving on the verse with another popular ditty:

> Great fleas have little fleas upon their backs to bite 'em,
> And little fleas have lesser fleas, and so ad infinitum.
> And the great fleas themselves, in turn, have greater fleas
>   to go on;
> While these again have greater still, and greater still, and so on.

How small are bacteria? A typical bacterium is about two microns in size, which means that about one thousand average bacteria could be lined up end to end on the head of a pin – or five thousand bacteria could line up end to end across your baby fingernail.

Until the later part of the nineteenth century, it was believed that bacteria were the smallest living organisms and the basis of all infectious diseases. There were certain niggling exceptions, however. Pasteur proved that rabies was infectious and even developed a vaccine for it, but was unable to isolate the rabies "germ."

In 1892, Russian plant scientist Dmitri Ivanovski tried to discover the cause of tobacco mosaic, a disease of tobacco plants. Using special filters with holes smaller than any known bacteria, he tried to isolate any bacteria – but failed repeatedly. It appeared that the germ responsible for the disease was very much smaller than any bacteria. Solvents such as alcohol or formalin, which usually killed bacteria, did not seem to affect this new "monster," as he called it.

As the nineteenth century was ending, scientists developed porcelain filters with uniform tiny holes or pores. Some of these filters had holes only 0.5 microns in diameter – one-quarter the size of the smallest bacteria. Any liquid that was passed through these filters would have all of the bacteria removed. The so-called Berkefeld filter eventually helped to solve the mystery of yellow fever.

The Yellow Fever Commission

In 1900, U.S. army surgeon Dr. Walter Reed (1851–1902) and English-born Canadian Dr. James Carroll (1854–1907) set up camp at Columbia Barracks in Quemados, about six miles from Havana, Cuba. Their mission: to find the cause of yellow fever. Yellow fever had plagued the southeastern United States for two hundred years, but nowhere was it more prevalent than in Havana. At seventeen years of age, Reed had been the youngest student ever graduated from the School of Medicine at the University of Virginia, and now headed the Yellow Fever Commission, with Carroll second in command.

Non-immune volunteers (who were each paid a $100 gold piece) agreed to be bitten by infected mosquitoes. These men developed yellow fever. Some were permanently damaged, and some even died – a risk they had agreed to when they signed on. But the experiment was deemed a great success because it showed that mosquitoes were indeed the vector for the disease. Still, they had not found the specific agent that caused the disease.

In summer 1901, Carroll took blood from yellow fever victims and passed it through one of the fine porcelain Berkefeld filters. Carroll then collected the resulting filtrate and injected small amounts of it into the body of a volunteer – who shortly thereafter developed yellow fever.

This proved that the yellow fever "germ" was filterable, or "ultra-microscopic," and therefore the "germs" that were causing the disease were not bacteria. But what were they?

What's Smaller than a Bacterium?

In 1933, a German named Ernst Ruska started experimenting with an electron microscope that he had invented and built himself. For the first time scientists had a microscope that gave better definition than a light microscope. Finally they could see these tiny "germs" that had so perplexed generations of scientists. What they saw amazed them – tiny, formed objects, some of which looked like beach balls, others like strings, and yet others like bizarre miniature machines.

In 1935, American biochemist Wendell Stanley made a remarkable discovery. He found that the "germ" causing the tobacco mosaic disease Ivanovski had studied in tobacco plants could be purified and made into crystals just like salt or sugar. These crystals, when dissolved, would once again cause disease. When these crystals were analyzed, they contained mostly protein with a small amount of DNA, or its close counterpart RNA. Stanley received (with two colleagues) the Nobel Prize for Chemistry in 1946 for his work in the purification and crystallization of viruses, thus demonstrating their molecular structure. (Ruska received the Nobel Prize for Physics in 1986, twelve years after his retirement.)

These tiny, infectious particles were named viruses from the Latin word for *slime*, particularly that which is foul or poisonous. Scientists already knew that bacteria, if just given some food such as broth, could move, eat, excrete, and reproduce. Viruses, however, display a much smaller range of abilities. Their single talent seems to be reproduction – and then only inside living things.

How do viruses cause disease? It takes but a single virus to get inside a cell. When inside the cell, the DNA in the virus takes over the cell's metabolism and causes the cell to manufacture more viruses. When the cell is full of virus particles it ruptures, and hundreds of newly formed viruses are released. Because viruses can neither live independently, nor feed nor excrete waste products, there is a lively debate within the scientific community about whether viruses can be classified as life. This debate recently took on a new perspective when in 2002 Dr. Eckard Wimmer reported that his team at the State University of New York at Stony Brook had assembled a "working" poliovirus from chemicals purchased from a mail-order supplier. After they assembled the virus, they injected it into a mouse, where it caused disease.[3] Although the scientific, moral, religious, and even military implications of this enterprise are enormous, the story received little notice in the press.

If a thousand average bacteria could be lined up end to end on a pinhead, and a porcelain filter can capture all bacteria, how small is a virus that can still slip through that filter? Jonathan Swift had the right idea: "And these have smaller still to bite 'em." Up to one thousand viruses could be lined up end to end – on a single bacterium. If you could get them to behave, you'd need 5 million viruses to form a straight line across your baby fingernail. Viruses are very small indeed.

KURU, SCRAPIE, AND CJD: ASSEMBLING THE PUZZLE

With the discovery of viruses, surely the entire stable of infectious agents had at last been identified? Infection could be spread by bacteria, viruses, fungi, and parasites, and nothing else. This erroneous belief persisted until the 1960s.

Still, there was that bizarre disease called scrapie that had plagued Scottish sheep since at least 1730.

In 1772, in one of the first recorded reports of scrapie, Reverend Thomas Comber referred to the disease as rickets, probably because the affected sheep demonstrated walking difficulties.

> The principal symptom of the first Stage of this Distemper, is a Kind of Light-Headedness, which makes the affected Sheep appear much wilder than usual, when his Master or Shepherd as well as a stranger, approaches him. He bounces up suddenly from his Laire, and runs to a Distance as though he were pursued by Dogs, . . .
>
> In the second Stage of the Distemper, the principal Symptom of the Sheep is his rubbing himself against Trees, Posts &c. with such Fury as to pull off his Wool and tear away his Flesh . . .

The third and last Stage of this dreadful Malady seems to be only the Progress of Dissolution, after an unfavourable Crisis. The poor Animal is condemned by Nature, appears stupid, separates from the flock, walks irregularly, generally lies, and eats little. These Symptoms increase in Degree till Death.[4]

This description bears an uncanny resemblance to both kuru and mad cow disease.

Reports of scrapie emerged sporadically in the United Kingdom for the next two hundred years. The disease appeared to vanish, then to crop up again. It is likely that the disease was elusive in part because farmers considered it shameful. The disease reflected poorly on the skills of the flock owner, and also considerably reduced the value of his flock. Farmers naturally took pains to conceal it.[5]

Because scrapie posed a threat to one of Britain's major industries of the time, it was extensively studied and debated. The main debate centred on whether the disease was infectious. Convinced that it was, a German writer in 1759 advised any farmer who had an infected sheep to slaughter it immediately and feed the meat to the servants – an unfortunate recommendation.

In the 1930s, British veterinarian Dr. William Gordon, in vaccinating sheep against another disease, louping-ill, inadvertently transmitted scrapie to the sheep. Although he had added formaldehyde to his inoculum to inactivate the louping-ill virus, that precaution failed to destroy the scrapie "virus" present in the inoculum. Formaldehyde killed all known viruses, so why not scrapie? Gordon's inadvertent error demonstrated a number of features of scrapie:

- The infective agent of scrapie was present in the brain, spinal cord, and/or spleen of infected sheep (from which he had made the inoculum).
- It could withstand a concentration of formalin (formaldehyde) of 0 to 35 per cent, which inactivated the virus of louping-ill.

- It could be transmitted by subcutaneous (under the skin) inoculation.
- It had an incubation period of two years and longer.

Kuru and Scrapie: Another Clue

1959. London, England. While working on scrapie at the Agricultural Research Field Station in Compton, England, William Hadlow, an American veterinary neuropathologist, visits a medical museum in London. There he sees an exhibit featuring photomicrographs of the diseased brains from a tribe in Papua New Guinea – probably taken from Ivan Klatzo's preparations of the material sent by Carleton Gajdusek. The brains of the kuru victims look very much like the brains of sheep that died from scrapie – full of spongelike holes. Hadlow notes "the vacuolated neurons, unusual in human brains, that were so much like those of scrapie. . . . I found the overall resemblance of kuru and scrapie to be uncanny."[6]

Kuru, CJD, scrapie – three different neurological diseases with many similarities. There had to be a connection.

By 1959, British veterinary researchers had conducted experiments to try to answer the question of scrapie infectiousness. In a typical experiment, they removed the brain and/or spinal cord from an animal that had died of scrapie. They then ground up this material, purified it, and either fed it to or injected it into another sheep. They then observed that sheep for up to two years to see if it developed the disease. After many years of study, these scientists discovered that scrapie could be transmitted through extracts injected from a diseased animal to a healthy one.

The very long incubation periods and relative high cost of the experimental animals made research difficult. In 1961, Professor Richard Chandler of Imperial College, London, made one of the most significant discoveries in the entire scrapie story. He discovered that scrapie could be transmitted to mice. Research could now be done in a lab using animals that were familiar to scientists and that were thousands of time cheaper than sheep. Furthermore, the incubation period for these diseases in mice, while long, is still much shorter than in sheep. From a medical point of view, Chandler's discovery also proved that whatever caused scrapie was capable of crossing species lines. This was a major finding, as most infectious diseases only attack one species – you won't catch kennel cough from your faithful Labrador retriever, but neither will she catch rubella from you. These diseases are said to be species specific.

Taking it one generation further, an extract from the second animal could cause the disease in a third animal. The line of transmission looked like this: Taking some brain material from a sheep with scrapie, Chandler injected it into a mouse. The mouse developed a spongiform encephalopathy. Chandler then took some of the brain from this infected mouse and injected it into another mouse. This second mouse also developed the disease. This evidence satisfied Koch's postulates, so the disease could be said to be infectious. The infectious agent, however, could not be isolated. As was the case with kuru, it was assumed to be a very small virus.

What's Smaller than a Virus?

In 1966, British radiopathologist Dr. Tikvah Alper and her colleagues showed that this "virus" was indeed very small. If a thousand bacteria could line up on the head of a pin, and a thousand viruses could line up on the back of a single bacterium, how much smaller could the scrapie agent be? Alper showed it to be the size of a single protein molecule – or about one one-hundredth the size of the smallest known virus.[7]

One year later she set the scientific world on its ear with an even more unusual finding. She found that the agent that caused scrapie contained neither RNA nor DNA.[8] This was tantamount to scientific heresy. Every infectious agent from parasites to bacteria – even viruses – contained these chemicals. Yet here was an agent that could replicate despite its lack of nucleic acid. Scientists could not begin to explain how an agent without RNA or DNA could cause an infectious disease. And since they could not explain her findings, they chose to ignore them.

10

CHAPTER

THE ULTIMATE STEALTH INVADER

Enter Dr. Stanley Prusiner. An ambitious American neurology resident at the University of California, San Francisco, Prusiner began looking into "slow virus" diseases in 1972 after admitting a female patient who was exhibiting progressive loss of memory and difficulty performing some routine tasks. He was surprised to learn that she was dying of Creutzfeldt-Jakob disease (CJD), a rare and little-known disease. Reviewing the literature, he found that kuru, scrapie, and CJD had all been transmissible by the injection of diseased brain extracts into healthy animals. No one had ever isolated the "slow-acting virus" thought to be the cause of infection.

Furthermore, even when experimenters had used all known procedures to denature the nucleic acids and inactivate viruses, the extracts from diseased brains remained infectious. However, when the extracts were exposed to a protein denaturing agent, guanidine hydrochloride (which does not kill DNA or RNA), the extracts were rendered uninfectious.

Prusiner, a neurologist and biochemist, set up a lab to study scrapie in 1974. Building on the previous work of numerous scientists, Prusiner and his team began a series of experiments to identify precisely the infectious agent. The search would take them ten years. He had anticipated that the purified scrapie agent would turn out to be a small virus and was puzzled when the data kept telling him their preparations contained protein but not nucleic acid.

To resolve the problem, first he took a portion of an extract from the brain of a scrapie-infected sheep and subjected it to chemicals that destroyed all nucleic acids (DNA and RNA). He found that an extract thus treated remained just as infectious in an experimental animal as one that had not been treated. This result confirmed radiation biologist Tikvah Alper's observation in the 1960s that the infectious agent did not contain these chemicals. Prusiner then took another portion of the same extract and subjected it to chemicals that left the DNA and RNA intact but destroyed any protein in the sample. This second sample did not cause disease in the inoculated animals. He concluded that the infectious agent must therefore be a protein.

Prusiner recounts the events that ensued:

> As the data for a protein and the absence of a nucleic acid in the scrapie agent accumulated, I grew more confident that my findings were not artifacts and decided to summarize that work in an article that was eventually published in the spring of 1982. Publication of this manuscript, in which I introduced the term "prion", set off a firestorm. Virologists were generally incredulous and some investigators working on scrapie and CJD were irate.[1]

Some of that incredulity and anger remains today. But the outcome would eventually see Prusiner vindicated on one of the world's biggest public stages.

PROTEINS, PRUSINER, AND PRIONS

Proteins are the basic building blocks of life. DNA contains the instructions for the body to manufacture proteins, which are made up of hundreds to thousands of simpler molecules called amino acids. The human body contains millions of different proteins. Protein makes up about 20 per cent of the body by weight. If all the water is removed from the body, the remaining material is about 60 per cent protein by weight.

For hundreds of years scientists have known that a molecule composed of more than one component part can assume different shapes, and these different shapes can have different properties. Think of water, ice, and steam. All are the same molecule but in different configurations. It is these different configurations that give water, ice, and steam their diverse physical properties.

After DNA (deoxyribonucleic acid), proteins can be thought of as the basic constituent of all living things. The nucleus of a cell contains mainly DNA, which stores genetic information for a cell. This is the blueprint for every protein that the body manufactures. The nuclear DNA sends instructions on how to assemble a protein to small structures inside the cell called ribosomes. These ribosomes assemble a protein from its constituent parts, which are smaller molecules called amino acids. These amino acids are added end to end until the entire protein is assembled.

Once the newly minted protein separates from the ribosome, it immediately folds into a complex shape that determines what function it will perform in the body. There are literally thousands of shapes any particular protein can assume, but only one correct one. Any deviation from the correct final form will result in a protein unable to perform its biological functions.

Until fairly recently, the proteins were thought to be self-assembling, which means that once formed, the protein would assume its correct shape unaided. This entire notion was challenged in the 1990s by a team of scientists in Germany studying a strange group of micro-organisms called Archaebacteria. (This term has since been abandoned, and the whole group of organisms is now called simply Archaea.) These very ancient organisms live in extreme conditions such as near rift vents in the deep sea, at temperatures well above the boiling point of water. Most proteins are very sensitive to high temperatures and cannot survive above 100 degrees Celsius. Since these Archaea are able to thrive, there must be some mechanism in them that prevents heat damage to the proteins.

The scientists discovered that the Archaea contain a cylindrical protein complex, one of a class of hardy proteins known as the "heat-shock" proteins. These proteins are released when a cell is in danger of suffering heat damage. The unusual talent of heat-shock proteins is their ability to cause proteins that are becoming misshapen due to the heat to regain their normal configuration. In other words, they return the organism's proteins to a correctly folded state, thereby preventing cell damage.[2]

In recent years, these same proteins have been found in normal cells not subjected to heat trauma. In fact, they are probably present in all cells. They have now been renamed "chaperone" proteins, or molecular chaperonins. Chaperone proteins appear to have at least two major roles in the cell. First, they serve as a kind of template to assure that newly formed proteins and large protein complexes assume the correct shape. Second, they help proteins that have folded wrongly as a result of stress to get back to the correct shape. This task plays an important role in normal bodily function. Due to minor genetic variations some individuals may manufacture proteins with one or more incorrect amino acids. As many as a third of proteins may end up misfolded. If these changes are not too great, the chaperone proteins can alter the shape of these incorrect proteins sufficiently enough to function in the body.[3]

These discoveries may help to explain why people with inherited disease such as muscular dystrophy appear fine early in life, then deteriorate as they get older. It's easy to imagine a situation where the chaperone proteins keep realigning the malformed muscle proteins, but as the patient ages and continues to manufacture defective proteins the chaperones eventually become overwhelmed, and malformed non-functioning proteins continue to accumulate, and increasing disability ensues.

This concept of misfolding proteins and chaperones is relatively new but is getting a great deal of attention in medical research, as it offers a good explanation for many poorly understood disease processes. The list of diseases caused by misfolded proteins continues to grow. Some of these "conformational diseases" as they are now called include Alzheimer's disease, Pick's disease, Huntington's chorea, and CJD. This model also offers new insights of great value in designing new therapeutic drugs.[4]

In 1982, Stanley Prusiner published his controversial manuscript in which he proposed that the infectious agent for scrapie and, by extension, for all TSEs was not a slow virus, but rather a "nucleic acid-free proteinaceous infectious molecule" – a rogue protein. PrP is a normally occurring protein found in the walls of cells. Prusiner's culprit was a misfolded protein – a copy of normal PrP that had "gone bad." Prusiner coined the word *prion* – derived from *pro*teinaceous *in*fectious particle – for this rogue protein. His detractors are fond of pointing out that the word should have been *proin*, but since Prusiner named it, *prion* has stuck.

His team also discovered that the gene encoding for PrP is found in all mammals, from mice to humans – so having a protein capable of becoming a prion is, in fact, normal. But the PrP protein has at least two forms. Prusiner discovered that the normal form, a tightly coiled helical structure, does not cause disease. But when the protein unwinds, it becomes a stable protein capable of causing infection. This new stable configuration of the PrP protein is now a prion.

By a mechanism that is still poorly understood, when one of these misfolded PrP proteins comes into contact with a normal PrP protein, it causes the normal protein to assume the abnormal shape, thereby becoming a prion. This new prion could induce other normal proteins to misfold and thereby set in motion a very slow chain reaction. (Science-fiction aficionados have noted the similarity to "Ice-9," a fictitious substance able to provide all water on earth with a different molecular structure, as described by Kurt Vonnegut in his 1963 novel *Cat's Cradle*. Such a substance could destroy the world.)

These prions have very different properties from the normal PrP proteins. Normal PrP is easily dissolved in water. Prions are not. Furthermore, prions are biologically inert, unable to take part in normal biological reactions. Prions tend to clump together in large aggregates that constitute the plaques present in the brains of victims of prion diseases. As more and more PrP proteins change into prions, the cells that contain these proteins die, leading to the symptoms found in victims of these diseases.

Prions are sometimes compared to beta amyloid protein, particles that form insoluble plaques in the brains of Alzheimer's disease victims. Prions fold into structures called beta-pleated sheets in a pattern similar to amyloid found in AD. Normally, our cellular machinery removes misfolded proteins. But when this process fails, misfolded proteins can begin to accumulate and stick together, forming tiny filaments and larger fibrils in the brain. These in turn aggregate into insoluble protein deposits called amyloid plaque – a distinctive feature seen at autopsy on AD victims.

You Can't Uncook an Egg

To understand how prions cause the changes in a brain (whether human, sheep, or other species), imagine a large bowl of raw egg whites. These whites are made up of a protein called albumin, which is folded in a unique way. Because of its shape, this protein is a translucent liquid and can be dissolved in water. Now place a tiny amount of the egg whites in boiling water, where it cooks immediately. The protein is still albumin, but now it is a solid instead of a liquid. It is now bright white rather than translucent, and it will not dissolve in water.

Most importantly, there is no way that the cooked egg whites can be made to revert to their uncooked state. All these changes happen because the protein is now folded in a different manner. The cooked form is said to be a more stable configuration of this protein, meaning this reaction can only go from raw to cooked, and never in the opposite direction.

If one were to take a teaspoon of cooked egg whites and chop it up very finely, that would be analogous to taking an extract from the brain of a sheep with scrapie. In our imaginary experiment, we would then mix the cooked, chopped egg whites with the raw egg whites. If the cooked egg whites behaved like prions, several months to years later the bowl of raw egg whites would be filled with large clumps of white material – hard-boiled egg white. The misfolded (i.e., cooked) protein causes the normal (i.e., raw) protein to misfold and these new prions cause others to misfold, thereby setting off an extremely slow chain reaction. An extract from this bowl, added to a second bowl of raw egg whites, would cause the same reaction. The prions appear to be in a form so stable they cannot be corrected by the action of chaperone proteins.

In another experiment, Prusiner asked the question, "Is the PrP protein necessary for life and, if it is not, can an animal that does not have this protein get a prion disease?" Using the tools of molecular biology he was able to "knock out" the gene responsible for making PrP in mice. The mice that were born without this gene did not have PrP proteins in their brains, but despite this seemed to be perfectly healthy, and lived a normal lifespan, which proved that PrP, at least in mice, is not essential for life. He also found that mice without this protein are resistant to contracting scrapie, even when inoculated with the prion that causes disease.

Thus the prion hypothesis was proved. Here at last was the vital connection – a rogue protein – that linked the neurological diseases kuru, scrapie, and CJD.

After having suffered professional skepticism and personal insult for his suggestion of a new infectious particle, Prusiner received the ultimate public vindication. He received the Nobel Prize in Physiology or Medicine in 1997.

HOW DANGEROUS ARE PRIONS?

If a thousand average bacteria could be lined up end to end on a pinhead, and up to a thousand viruses could be lined up end to end on a single bacterium, what's "smaller still to bite 'em"? Between one hundred and one thousand prions can fit into a single virus.

Not only are they small, but prions are exceedingly hard to destroy. They are very resistant to heat – several patients have contracted prion diseases from surgical instruments that had been autoclaved (heat sterilized). Agents that kill almost all bacteria and viruses, such as ultraviolet or ionizing radiation and solvents such as alcohols or formaldehyde, do not inactivate prions. It's no wonder that surgeons take extreme precautions when performing neurosurgery on individuals who may be suffering from CJD. The U.K. Advisory Committee on Dangerous Pathogens, Spongiform Encephalopathy Advisory Committee strongly recommends that hospitals destroy by incineration all instruments and protective clothing used during surgery on such patients. Coroners and pathologists also show great respect for the dangerous infectivity of prions when dealing with CJD victims.

Almost every tissue from an infected animal is infective, but the most infective is neural tissue such as brain and spinal cord, followed by lymphoid tissue, such as spleen and tonsils. However, meat, blood, and even feces and urine in some studies have proved infective. The degree of infectivity in tissues such as brain is staggering. In experimental animals, one gram (about one-thirtieth of an ounce) of brain contains an amount of prions sufficient to infect well over one billion individuals.

What is more worrying is that animals that show absolutely no sign of disease can still be highly infective. In fact, because of the long incubation period of prion diseases, animals with short lifespans (or those that are slaughtered early in their natural lifespan) may never show signs of the disease, but can be highly infectious to susceptible species that consume their remains.

The natural lifespan of an unconfined chicken can be up to ten years, yet chickens raised for meat are usually slaughtered at six weeks of age, before they reach sexual maturity. Egg-laying hens are kept for two to four years. The natural lifespan of pigs is about ten to fifteen years, and even up to thirty years, but they are usually slaughtered after four to seven months. Pigs intended for pork are usually slaughtered one to two months younger than pigs for bacon. Sows kept as breeding stock are slaughtered after three or four years for sausages, pork pies, and other lower-quality products. These animals that never reach their full, natural lifespan may act as silent reservoirs for the disease.[5]

In 1999, scientists reported that they had observed beta amyloid plaques in the brain of a sixteen-year-old woodpecker. Although the woodpecker plaques tested negative for prions, this is the first reported example of these types of lesions in a bird.[6] This finding may mean that birds are capable of getting these types of diseases, but because chickens, turkeys, geese, and ducks are slaughtered so young in the food-processing industry, they do not live long enough to demonstrate any symptoms or pathological findings. Products from the poultry industry constitute one of the largest inputs to the rendering industry – and it's still legal to feed poultry products to cattle.

CROSSING THE SPECIES BARRIER

Prior to the BSE outbreak, common belief held that there was an entire family of prions just as there is a large family of viruses. These prions, like their virus counterparts, were presumed to be species specific. People cannot contract hoof-and-mouth from eating infected cattle because the virus causing this disease only infects cattle. Humans cannot get scrapie (so it seemed) because the prion causing this was also species specific. This idea of species-specific infections pervades both human and veterinary medicine. Agents that cause the same or similar diseases in different species are the exception rather than the rule. As anyone who owns pets will affirm, dogs may get mange or distemper, but their owners never do. Similarly a cat will not contract measles or pneumonia from its owner.

There had been a few reports in the United States of people with CJD who had eaten deer, elk, or in a few cases squirrels (particularly squirrel brains in a stew recipe called "burgoo," much favoured in Kentucky and other parts of the southern U.S.). But again, because these cases represented such a small number of people, no one made the link between eating these wild animals and contracting the disease.

Once prions were recognized as a cause of disease, similar diseases – all transmissible spongiform encephalopathies (TSEs) – were identified in other species. There were forms that affected deer and elk (chronic wasting disease, or CWD), squirrels, mink (transmissible mink encephalopathy, or TME), and others.

TME was first detected in the U.S. in 1947. Since then, TME outbreaks have been reported in numerous locations worldwide, including Canada, the United States, Finland, Germany, and the republics of the former Soviet Union.

In 1985, an outbreak of TME in Stetsonville, Wisconsin, lasted for five months. Clinical signs in the mink included tail arching, incoordination, and hyperexcitability. Eventually the affected animal would enter a trancelike state and die shortly thereafter. In the herd of 7,300 adult mink, 60 per cent of the animals died. Microscopic examination of sections of the mink brains showed the spongelike holes in the brain characteristic of TME and other TSEs. Diagnostic tests identified the prion protein.

On this particular ranch, the mink's diet had consisted of fresh meat products from "downer" cattle and commercial sources of fish, poultry, and cereal. Downer cattle are those that fall down and can't get up due to a condition such as a metabolic disease, broken limbs, or a central nervous system disorder. They may also be cows that die from some unknown cause and are considered unfit for human consumption.

The late Richard Marsh, a veterinary virologist at the University of Wisconsin who studied the transmission of TME and other TSEs, investigated the Stetsonville TME outbreak. He theorized that the meat from the downer cattle introduced a TSE agent to the mink, leading to the outbreak of TME.

In this case an extract from the brains of the mink that had died was injected into the brains of two Holstein steers. Both of these animals developed a prion disease, confirmed at autopsy. So the chain of transmission looks like this: Downer cattle meat $\rightarrow$ mink. Mink develop TME, a prion disease. Mink brain extract $\rightarrow$ into cattle. Cattle develop a prion disease.[7]

Despite the numerous clues from these kinds of reports, when the mad cow (BSE) epidemic hit the United Kingdom, it took ten years before authorities concluded that cannibalism had caused the epidemic. Cows that had died and were considered unfit for human consumption were made into animal feed. Most of this product went to feed other cows, but some was made into pet food for cats and mink.

Five years after mad cow disease showed up in England, on May 10, 1990, it was learned that a five-year-old male Siamese cat had died of a spongiform encephalopathy – the first known case of feline spongiform encephalopathy (FSE). This resulted in a rash of media comment, speculating that the cat had caught BSE – mad cow disease – and that humans might be next. Other reports of cats with FSE, including domestic cats as well as a puma and a cheetah, soon followed.

Both of these veterinary outbreaks and outbreaks among zoo animals seemed to point to infected food. Cats don't eat other cats and mink don't eat other mink – but both of these species eat cattle in the form of rendered animal feed.

In a rather strange case reported in the *Lancet*, doctors in Italy reported the misfortune of a sixty-year-old man who contracted CJD in November 1993. At the same time, his spayed seven-year-old female shorthaired cat came down with the feline version of the disease.[8] The cat was usually fed canned food and slept on its owner's bed. No bites from the cat were recalled. Both the cat and its owner died in January 1994. This was the only such case ever reported. Although it may be interesting to speculate on the dietary habits of both victims, most regarded this case as a curiosity news item.

As these veterinary outbreaks were fairly small and localized, they did not attract much attention from the scientific community, and because people eat neither cat, nor mink, nor zoo animals, they did not appear to represent a public-health threat. However, the dogma about species-specific prion diseases was becoming seriously eroded.

Brain extracts from BSE cattle have been shown to cause disease in cattle, sheep, mice, pigs, and mink when injected into the brains of those animals. The list of animal species that have been shown to be susceptible to TSEs transmitted from other species is, in fact, quite lengthy, and includes the following:

Human	Rat
Sheep	Hamster
Mink	Gerbil
Cow	Vole Guinea Pig
Deer	Rabbit
Chimpanzee	Pig
Gibbon	Puma
New-world monkeys (five species)	Cheetah
Old-world monkeys (eleven species)	Kudu
Goat	Nyala
Ferret	Gemsbok
Cat	Eland
Dog	Oryx
Raccoon	Moufflon
Skunk	Civet[9]
Mouse	

However, until the 1990s scientists still maintained that the prion responsible for TSEs was species specific. A rare diseased animal caused the mad cow epidemic and other animals of the same species ate the infected meat and got the disease. The cats and mink were considered some form of anomaly and were not considered as exceptions to the species-specific rule. This species-specific dictum was believed to be true until the case of Alison Thorpe.

Very recent research from a group in Germany has shown that fish also suffer from a prion disease.[10] This finding is of great significance, as products from aquaculture operations can be used in rendering plants, and the feed used in these operations may contain product from rendering plants. These results may cause a rethinking of the safety of fish raised in aquaculture operations where one of the main sources of food is the remains of other fish. This is analogous to the practice of feeding cows to cows – and we know where that led.

INCREASED VIRULENCE

The transmission of prions from one species to another usually requires a longer incubation time compared to transmissions where the host species is the same. This longer incubation period can lead to massive epidemics when infective material is consumed widely before symptoms show up.

Prions seem to act faster in successive generations. Hamsters were injected with brain extract from the Stetsonville mink identified by virologist Richard Marsh. The first hamsters to receive the mink TME died before contracting the disease, which took hundreds of days to incubate. The research team then injected brain material from those hamsters into a second generation of hamsters. The disease struck faster. Then they did it again to a third generation, and found a tremendous shortening of the incubation time.

Why did the prions become more virulent? It appears that upon interspecies transmission, two hamster-adapted TME strains emerged from a single mink TME strain. The long-incubation prion predominated in the first passage, but in succeeding generations the faster-acting prion grew dominant.[11]

ARE PRIONS ALL BAD?

Do prions do anything positive in either humans or animals? Scientists have discovered that prions may play a major role in the evolutionary process. They do this by affecting the proteins that control the expression of genes. This phenomenon has been directly observed in yeast cells, but is suspected to occur in other organisms as well.

It seems that cells have a built-in genetic variation that never surfaces until a new or changing environment threatens their survival. When that happens, the prion converts the previously neutral genetic variation to a non-neutral state. In other words, the prion allows yeast cells to exploit pre-existing genetic variation in order to thrive in fluctuating environments – a trick that may assist the evolution of new traits.[12] (Others disagree with the interpretation of experimental data. They propose that the variability in traits is a side effect of disrupted gene expression, not an adaptation to facilitate evolution.)

In what may prove to be one of the great ironies in nature, Dr. Susan Lindquist and Dr. Eric Kandel have recently discovered a mechanism by which long-term memories are stored in the brain. The mechanism involves a protein called CPEB. When long-term memories are laid down this protein alters its shape and causes other proteins to do the same thing. In other words it acts as a prion. This suggests that the same mechanism that is responsible for creating long-term memory is also responsible for destroying it.

Dr. Lindquist notes, "This is remarkable not just because the protein executes a positive function in its prion-like state. It also indicates that prions aren't just oddballs of nature but might participate in fundamental processes."[13]

In 2002, the United States provided $42.5 million to the Department of Defense to establish the National Prion Research Program (NPRP). The goal of the project is to develop rapidly a diagnostic test to detect the presence of prion disease. Such a test would be used as an antemortem test for TSEs, particularly in beef. But BSE is not the only concern. Chronic wasting disease (CWD), the prion disease of elk, is a huge and growing problem in the U.S. It's noteworthy that the funding goes through the military, a clear indication that the government believes prion diseases pose a significant threat to the health and blood supply of the country.

CHAPTER 11

OTHER HUMAN PRION DISEASES

Prion diseases are fatal neurodegenerative maladies that may present as sporadic, infectious, or genetic illnesses. The sporadic form of human prion disease is Creutzfeldt-Jakob disease (CJD), while the infectious forms include kuru and variant Creutzfeldt-Jakob disease (vCJD). In the last fifty years, other prion diseases have been identified in humans. The genetic or inherited prion diseases are called fatal familial insomnia (FFI), Gerstmann-Sträussler-Scheinker (GSS) disease, and familial (f)CJD.

FATAL FAMILIAL INSOMNIA (FFI)

> Methought I heard a voice cry "Sleep no more!
> Macbeth does murder sleep," the innocent sleep,
> Sleep that knits up the ravell'd sleeve of care,
> The death of each day's life, sore labour's bath,
> Balm of hurt minds, great nature's second course,
> Chief nourisher in life's feast, . . . – *Macbeth*, Act 2, Scene 2

Anyone who has ever suffered insomnia may be chastened by the following true story (some details have been changed). Like Macbeth, Mr. P does not sleep – he cannot sleep.

Early 1974. Mr. P, a normal, hard-working thirty-three-year-old executive living in a modest suburb in New England, begins to experience difficulties sleeping. He finds it almost impossible to fall asleep, and if he does manage to drift off on rare occasions, he cannot stay asleep. He lies in bed staring at the ceiling, reading books, watching late-night television – anything but sleeping.

Unlike other people with sleep disorders, he doesn't fall asleep during the day nor does he have little "catnaps."

After more than a month, he starts to have hallucinations – both visual and auditory. He sees people who are not there. He hears sounds or conversations that no one else can hear.

At about the same time, he begins to experience tics and involuntary muscle movements, almost as if he were being frequently startled. Both he and his doctor feel the hallucinations and tics are probably secondary to the sleep deprivation.

Mr. P tries everything to get to sleep – Valium, barbiturates, and other sleep-inducing compounds. These medications make him feel groggy and stupid, but don't put him to sleep. Alcohol has a similar effect. Hypnosis, biofeedback, and counting sheep are similarly ineffective. Finally, after three months of almost no sleep, a forty-pound weight loss, increasingly frequent hallucinations, and tics and spasms, Mr. P is admitted to hospital.

The doctors who admit him note he looks acutely ill. His blood pressure and temperature are both elevated. He has a fever. He is disoriented: he doesn't know what day it is or where he is. Sometimes if asked a question, he makes up an answer that is obviously untrue. For example, when asked where he is, he might answer, "At a restaurant."

Now Mr. P is shaking uncontrollably and having major jerking motions of his arms and legs. At first, doctors think he might have meningitis, so they place a needle in his spine and draw out some cerebral spinal fluid for analysis. It is completely normal. Blood tests are done to check for any infection or dysfunction in any organ or gland. All these tests are normal. Finally, doctors order every blood test imaginable – all come back normal.

X-rays and CT scans – normal. An EEG is done to measure brain activity – normal. Finally out of desperation, doctors take Mr. P to the operating room. They drill a hole through his skull and remove a small piece of the exposed brain for analysis. The result of this brain biopsy – normal.

Meanwhile, Mr. P is deteriorating rapidly. His insomnia continues, as do the weight loss and the hallucinations. The twitches get worse. He becomes completely demented. His eyes jerk back and forth like a strange metronome. His eyeballs appear to be bulging out of their sockets. He is unable to walk. He experiences increasing difficulties with swallowing and breathing. Finally, he dies. The time from the onset of his insomnia to death: nine months.

A frightening tale indeed. What was wrong with Mr. P? Studies done at autopsy and genetic studies revealed several somewhat abnormal findings in his brain, and a small mutation in two genes, but nothing that would identify the disease that had killed him.

Five years later, fifty-seven-year-old Mr. P's aunt (his mother's sister) began stumbling. Her speech became slurred, and she lost twelve pounds, for no apparent reason. Like Mr. P, she also took a rapid downhill course, becoming completely demented. About eighteen months after the first onset of symptoms, Mr. P's aunt died. Interestingly, she did not complain of insomnia at any point in the disease.

Two years later, Mr. P's mother experienced difficulties with walking, and she too rapidly deteriorated in a manner similar to her sister and her son. She died twenty months following the onset of her symptoms.

Several years later, Mr. P's mother's cousin also succumbed to a similar condition.

After doing DNA studies in all these cases, researchers found a mutation on a specific gene, which seemed to be the reason this family was susceptible to this terrible condition.

This particular family had an extremely well-documented family tree going back to the thirteenth century in England. A branch of the family immigrated to the United States arriving only fifteen years after the *Mayflower*.

Poring over family records, researchers found that the first case of a similar-sounding condition occurred in 1894, with a family member dying at the age of forty from a disease characterized by "muscle twitching." Three of his four children had died at ages forty, fifty, and fifty-two.

All their death certificates indicated they had succumbed to a neuro-degenerative disease.[1]

At first, researchers thought this unfortunate family suffered from a condition so rare they might be the only family in the world so afflicted. As this disease had never been described before, it was considered to be a new condition and was given the somewhat frightening name fatal familial insomnia (FFI).

Since that time, however, families with similar history have been found in the United States, the United Kingdom, Austria, Italy, France, Germany, Australia, and Japan, and more cases are reported on a regular basis. In addition to cases linked to families, there have been many sporadic cases found that seem to have neither a family history nor any genetic defect.

To date, twenty-four families have been identified worldwide with the mutation that makes it more likely they will contract the disease. Insomnia, however, appears to be a less than constant feature, so some experts have debated whether FFI is the best possible name for the disease, although it is well entrenched in biomedical terminology. Some cases are more notable for rapid weight loss, leading Herbert Budka from the Institute of Neurology, University of Vienna, to suggest that a term such as *neuroemaciation* might reflect this most useful and striking early sign. However, many signs and symptoms considered characteristic for FFI may develop in CJD as well. About 10 to 15 per cent of patients with CJD report sleep disturbances among their early symptoms. In rare patients, this insomnia can mimic the intensity of FFI.

In 1995, FFI was also shown to be transmissible. Researchers injected extracts from the brains of people who had died of the disease into experimental animals, whereupon the animals developed a form of the disease.[2]

FFI is certainly dramatic in its presentation; however, except for the insomnia, it is almost exactly the same as CJD. It is likely that FFI is a variant of CJD. Autopsies show the spongiform changes in the brain typical of prion diseases. There is no known cure for FFI. The only hope rests on gene therapy, which to date has been unsuccessful.

GERSTMANN-STRÄUSSLER-SCHEINKER DISEASE (GSS)

Gerstmann-Sträussler-Scheinker disease (GSS), another extremely rare, neurodegenerative brain disorder, greatly resembles CJD in its presentation. Almost always inherited, GSS is due to a defective gene encoding the prion protein, and is found in only a few families around the world – but with a surprising number of variations, given its rarity.

Like FFI, onset of GSS usually strikes people between the ages of thirty-five and fifty-five. Early symptoms may include clumsiness and difficulty walking. As the disease progresses, muscle coordination decreases, and patients may develop slurring of speech, involuntary eye movements, dementia, and sometimes blindness and/or deafness. In some families, Parkinsonian features are present. Death occurs anywhere between one and eleven years after onset.

On autopsy, the brains of some GSS victims show not only the spongiform degeneration found in other prion diseases but also particular multicentric amyloid plaques and neurofibrillar degeneration in the brain – identical to the markers found in the brains of Alzheimer's disease victims.

Like other prion diseases, there is no cure for GSS, nor are there any known treatments to slow progression of the disease.

HOW RARE ARE HUMAN PRION DISEASES?

All of these human prion diseases present similar clinical symptoms, while the findings at autopsy all show the distinctive plaques in the brains. It seems most likely that all these diseases – kuru, CJD, VCJD, FFI, and GSS – are slightly different manifestations of the same process and are probably, for all intents and purposes, the same disease.

Countries that employ surveillance programs for prion diseases include Australia, Austria, Canada, France, Germany, Italy, Netherlands, Slovakia, Spain, Switzerland, United Kingdom, and the United States. In these countries, total deaths from definite and probable cases of CJD (sporadic, familial, and iatrogenic) stood at 744 for 2001 and 633 for 2002, the last year for which complete figures are available.[3] This gives a rate of a little more than one in a million, which is fairly consistent from country to country – except for the United States, where CJD is not a reportable disease. Remember too that many neurodegenerative diseases are often misdiagnosed, which may skew the figures.

University of Michigan neurologist Dr. Norman Foster notes the case of two young men in Michigan who died of CJD in 2002. Both men were younger than thirty years of age. Foster believes it is possible that the odds might be much lower than previously thought for CJD in the young, and that many cases may be misdiagnosed. Although autopsies and other tests indicate the victims died from so-called "classic" forms of Creutzfeldt-Jakob disease, the two men did not have some of the typical symptoms of CJD. Their brainwaves were different and they lacked a particular diagnostic protein in spinal fluid. One was originally thought to have epilepsy. Foster suggested that many cases diagnosed as viral encephalitis might be CJD.

The incident raised fears that the human form of mad cow disease, or something similar, had emerged in the United States, but Foster emphasized these were *not* cases of variant Creutzfeldt-Jakob disease (VCJD). Entirely different prion agents caused both cases. They were not contagious.

Investigators from the U.S. Centers for Disease Control (CDC) looked for causes in the highly unusual and coincidental deaths (the men did not know each other, but lived in adjoining counties). The CDC ruled out foreign travel, eating deer or elk meat, and receiving human growth hormone. The forms of CJD suffered by the young men are ones seen previously in older individuals.[4]

Prion diseases may be much more prevalent than current official figures would have us believe.

Prion diseases affect primarily the nervous system. They can be sporadic (spontaneous), familial (inherited), or transmitted by infection. Over the last half-century, scientists working in far-flung corners of the globe, initially in isolation from one another, tried to discover what caused these ghastly neurological diseases. Gradually pieces of the puzzle began to fall into place.

In his groundbreaking work in figuring out what prions are and how they impart disease, Dr. Stanley Prusiner made some important discoveries.

- First, injecting prions is many times more likely to cause disease than is eating them. One scientific paper estimated that a single dose injected intracerebrally (i.e., into the brain) was equivalent to twenty-five thousand doses taken orally.[5]
- Second, ingesting (eating) prions is an extremely inefficient way of contracting the disease, which may explain the low incidence (one in a million) of the "sporadic" disease. This means that even if many thousands of people are exposed to prion-contaminated food in their diet, very few will actually contract the disease.

The low infectious capability of oral doses might explain the low native rate of the disease, but it does not explain why CJD appeared in the early 1900s in Europe, later in the United States, and why it did not occur in Japan until the 1960s. Prusiner also notes that the fact that CJD often seems to appear in clusters may be due to high doses of prion-laden material. "The occurrence of clusters of CJD patients could be due to ingestion of large doses of the CJD or possibly scrapie agents."[6]

COULD ALZHEIMER'S DISEASE BE A PRION DISEASE?

Both the behavioural and the pathological similarities between Alzheimer's disease (AD) and other prion-mediated diseases are so strong that it appears obvious they must be connected.

Remember that human prion diseases include CJD, VCJD, kuru, fatal familial insomnia (FFI), and Gerstmann-Sträussler-Scheinker (GSS) disease. Animal prion diseases include scrapie (sheep), BSE (mad cow disease), TME (mink), CWD (elk), FSE (felines), and many others.

So what about Alzheimer's? What is the evidence so far that might support this hypothesis that AD is a prion disease?

- In all prion diseases, the principal organ affected is the brain. The same is true with AD.
- All prion diseases cause dense amorphous deposits in the brain. AD also causes dense amorphous deposits in the brain.

- Dementia is the major symptom in all the human prion diseases. This is also true of AD.
- Because of their long incubation, in almost all cases of "naturally occurring" prion diseases the symptoms do not occur until later in life. This is also the case in AD. Furthermore, like many prion diseases, although it is obvious that the disease process has been going on for many years, there are no indications that the patient is incubating either disease until quite late in the disease process.
- There appears to be no way to reverse the disease once symptoms manifest. Like victims of other prion diseases, once diagnosed, the condition of AD patients deteriorates rapidly until the victim dies.
- Prion diseases were first described in the medical literature around the same time as AD (early twentieth century).
- In countries reporting CJD, AD is also reported. In countries such as India, both CJD and AD are exceedingly rare.

A great deal more compelling evidence suggests that AD, if not the same as other prion diseases, has a common antecedent – exposure to prion-tainted material in the diet.

PART FOUR

Eating Dangerously

12

CHAPTER

HOW NOW GROUND COW

For thousands of years humans have enjoyed an extraordinary relationship with the cow – a relationship both religious and secular. The earliest known examples of painting, the famous drawings in the caves at Lascaux and other areas in France, highlight this relationship. These paleolithic (early Stone Age) paintings, *circa* 15,000 B.C.E., depict many animals, but the animal most often shown is the aurochs (plural, aurochsen), an extinct wild ox and ancestor to our modern domestic cow. (The last individual aurochs was killed in Poland in 1627.)

The paintings underscore the dual relationship early humans enjoyed with these animals. Some artists depicted the aurochsen as objects of worship. Others show the beasts lying on the ground pierced with spears – clearly showing that humans could dominate these creatures. The large bulls had long, forward-pointing horns, and the smaller females shorter horns. The rampant masculinity of the wild bull together with the obvious femininity of the cow in the paintings suggests they stood as fertility symbols for ancient cultures.

Oxen, already domesticated by the neolithic (later Stone Age) period, are mentioned in the oldest written records of both the Hebrew and Hindu peoples, and have long represented a measure of wealth amongst those who owned the beasts. The image of an ox appears on some early Greek coins to signify its value as a trading instrument.

Ancient Egyptians worshipped cattle, particularly in the form of their god Apis, the god of strength in war and fertility. The powerful bull symbolized the strong and authoritative personality of the god. In the ancient city of Memphis, Egyptians kept a live bull as the incarnation of Apis, feeding it the best foods and housing it in a majestic stable. When the bull died, the new Apis was transported to Memphis on a boat with a specially built golden cabin. By Biblical times, cattle-worshipping cults were well established all through the Middle East and Western Europe.

The word *cattle* comes from the Old French *catel* or *chatel*, which became our modern word *chattel* – a movable possession. Those words in turn come from the Latin *capitale*, meaning holdings or funds. Thus do cattle go hand in hand with wealth.

"A MOTHER TO MILLIONS"

Cows are still regarded as sacred animals in India. Although they are seldom eaten, cows play a central role in day-to-day life. A quarter of the world's cattle population – more than 200 million cows – live in India. In most Indian states, it is illegal to kill cows.

Not only are cows objects of worship in India, but they also meet the dairy needs of the country. The ox is used in India as the tractor is in North America, providing the only non-human aid to agriculture for the majority of farms. Manure is one of the great renewable resources of the country. Half is used as fertilizer to replace soil nutrients. The other half is used for cooking fuel in the tandoors, or clay ovens, used throughout the country. In his book *Cows, Pigs, Wars, and Witches: The Riddles of Culture*, Marvin Harris estimates that using cow dung as fuel saves India on an annual basis the equivalent of 27 million tons of fuel oil, or 35 million tons of coal, or 68 million tons of wood.[1] It's no wonder that Gandhi called the cow "a mother to millions of Indians."

Britons have also worshipped cattle for thousands of years, but in an altogether different manner. The Celts had worshipped cattle in the years before the Roman conquest. When the Romans occupied Britain, they brought with them their own cattle culture. During the Roman occupation, beef was the preferred ration for the Roman army.

After the Romans left Britain, cattle remained the main measure of wealth. The eating of beef became synonymous with being British gentry. In medieval times, British nobility spared no expense in trying to outdo each other with lavish meat-based banquets. Although the rich ate beef almost to the exclusion of everything else, the poor could not afford such a diet. They had to eat "white meats," mainly cheese, milk, and butter. This diet also relied heavily on cattle.

In the 1700s, with the emergence of an industrial middle class, the demand for beef soared. In 1726, more than one hundred thousand head of cattle were slaughtered annually in London alone. In the eighteenth century, British sailors received an annual ration of 208 pounds of beef. It was believed that eating beef made men more fierce and better fighters.[2]

Even today, the Yeomen of the Guard, the British sovereign's personal bodyguards, are known as "beefeaters," a moniker given to them by the Grand Duke of Tuscany in 1669, who noted the hefty size of the Yeoman warders at the Tower of London. Every advertisement for British tourism features at least one beefeater wearing the well-known traditional uniform.

As Britain became more wealthy and powerful, the demand for beef grew. The rich landowners took to breeding bigger and bigger cattle. This race to see who could breed the largest, fattest cows became something of a national obsession among the aristocracy, culminating in 1800 with a national tour of the Durham Ox. This enormously obese animal weighed in at almost three thousand pounds. For six years he was paraded all over the British Isles in a specially built cart (shades of the Egyptian god Apis). People paid an admission fee to enter the tent and stare at his amazing bulk. This animal became the stuff of legends, and still today in Britain there are hotels, inns, and pubs scattered all over the country named The Durham Ox.

As symbols of conspicuous consumption, fat cows proved that the wealthy beef barons not only ruled over nature but also occupied a superior position in rural society, well above their peasant neighbours. It was also during this period that owners of country estates commissioned livestock portraits for their homes. In 1802, enterprising dealers sold more than two thousand prints of The Durham Ox alone, along with thousands of prints of other obese livestock. Today in English country homes, one can still see these vaguely silly pictures of enormously fat cows dominating paintings set in idyllic, bucolic surroundings. If fat cattle were a sign of wealth, then the eating of fatty or well-marbled beef was considered the ultimate in gourmet fare. This period marked the beginnings of Britain's passion for fatty beef.

THE FAT OF THE LAND

Edward Winslow, founder (later governor) of Plymouth Colony, had introduced cattle into New England in 1624, but cattle served only local needs for the next few hundred years.

By the late 1800s, as Britain's demand for beef continued to grow, the search for more pastureland drove British expansionism and foreign policy, and became one of the major factors in the opening of the American West. British capital financed the building of America's transcontinental railroads. British and Scottish bankers purchased vast tracts of land. Entrepreneurs spent huge sums of money to establish companies such as the Anglo-American Cattle Company – all this to supply Britain's insatiable demand for beef.[3] But this new method of ranching resulted in a problem: range- or grass-fed cattle produce lean meat, and the British (and American) consumers preferred their beef to be marbled – that is, fatty.

This dilemma led to one of the greatest industrial amalgamations of all time. Cattle companies realized that if range-fed cattle were given a diet rich in grain while at the same time kept relatively inactive, they would become delightfully fat. Thus was born the feedlot system. Cattle from the vast western plains were shipped by rail to the American midwest where they were fattened on grain from the midwestern grain belt. This was capitalism at its finest. Farmers could grow more grain than the country could possibly eat, and this excess was fed to cattle that had been shipped to the midwest using the newly developed rail lines, largely owned by British interests.

Toward the end of the nineteenth century, the beef-processing industry in the United States became a major exporter of American beef to the rest of the world, and especially to Great Britain.[4]

Two great inventions of the industrial age allowed the meat-packing industry to become one of the most important segments of the U.S. economy. These inventions were refrigeration and the assembly line – or, in the case of slaughterhouses, the disassembly line.

Ice Follies

In ancient times, Hebrews, Greeks, and Romans collected snow and ice and placed it in pits lined with wood or straw so that it would be available when the temperature became warmer. As early as 1000 B.C.E., the Chinese harvested and stored ice, a custom emulated by the French courts in an effort to keep its kings and queens cool in the middle of summer. The practice of trying to preserve ice and use it in the warmer months spawned an entirely new industry in the United States. The ice-house industry reached its pinnacle in the mid-1800s with a Bostonian, Frederic Tudor, who became known as the Ice King.

In 1875, *Scribner's Monthly Magazine* celebrated Tudor's remarkable career, noting that ice, "an article, once regarded as a simple luxury in non-producing countries, and in the northern latitudes as an article of no computed practical value, has become recognized in the commerce of the world."[5]

In 1805, yellow fever broke out in the British possessions in the West Indies. For whatever reasons – perhaps they wanted ice to relieve the suffering caused by the fever – this epidemic increased the demand for ice in these islands. Tudor cut about two hundred tons of ice from a pond on his plantation and trucked it to Charlestown. From there, it was loaded on a brig and shipped to Martinique. Most people regarded this as an act of great folly, but Tudor pressed ahead. Within a few years he had been granted a monopoly, and his ice houses were soon found on every major island in the West Indies. Customers arrived at the shipping docks carrying blankets in which to wrap their blocks of ice – or Tudor would sell you a blanket "of sufficient size" for a dollar.

From there he branched out both to domestic and foreign markets. In 1833, he took ice to India in a ship named *Clipper Tuscany* – the first time ice had been shipped into India from abroad. To expedite a more efficient harvest, Tudor invented a horsedrawn rake that marked the surface of a lake into twenty-inch rows. Workers then sawed the ice into sixteen-foot lengths that they stored in insulated ice houses. By the late 1860s, Tudor and other entrepreneurs like him were harvesting more than 200 million tons of ice per year in the United States. New York City alone consumed more than 1 million tons of ice per year.

Although Tudor struggled in the early years, even spending time in prison as a debtor, he eventually succeeded in convincing much of the world that cool beverages tasted better than warm ones, and that everyone living in even the warmest climates should be able to enjoy ice cream. The cost of ice to consumers ranged from five to twelve dollars a ton, or ten cents a pound in small quantities. Although the ice industry required a certain amount of labour and materials, the basic raw material – frozen lakes – still cost nothing. No wonder Tudor became a very wealthy man.

By 1900, the ice industry was in full swing. Once again this young country, America, was proving to be a capitalist's dream. Ice, like grass, was free for the taking, and could be sold for a tidy profit. In 1907, 14 million to 15 million tons of ice a day were consumed. This was a resource that could provide income for anyone with access to fresh water. In fact, entrepreneurs even harvested ice from Thoreau's famous Walden Pond, to the tune of one thousand tons of ice per day in the winter of 1847.[6]

But the days of the natural ice industry were numbered. In the last decades of the nineteenth century, natural ice became a public-health hazard. The practice of dumping industrial waste and raw sewage into the closest available body of water (a practice still not stamped out today) led to two results. First, harvesters had to travel longer distances to get clean ice, so the price became higher, and second, the public began to mistrust the safety of natural ice.

The Big Chill

The death of the natural ice industry gave birth to the refrigeration industry. Thomas Moore, a Maryland farmer, first coined the term *refrigerator* in 1803. Today, Moore's appliance – a cedar tub, insulated with rabbit fur, filled with ice, surrounding a sheet metal container – would be called an icebox. Moore designed his refrigerator for transporting butter from rural Maryland to Washington, D.C.

The first commercial use of refrigeration was in the brewing industry. Americans developed a preference for cold German lager over warm British ale. The large German immigration to the midwest in the 1840s further fuelled the demand.

In 1870, the S. Liebmann's Sons Brewing Company installed the first commercial refrigeration unit in the United States. This new technology caught on quickly in the brewing industry, and by 1891 almost every brewery had a refrigeration unit.

Ten years later, the meat-packing industry recognized the enormous potential of this new technology. With refrigerated storage facilities, meat could be stored year-round, not only in the winter. The real benefit of this technology was that it could be used in trains and ships, and soon thereafter even in trucks. This development allowed fresh (rather than canned or salted) meat to be shipped from the midwest to anywhere in the world.[7]

Now that they had the means to ship fresh meat, the meat-packing industry could begin to consolidate their interests – and their cattle.

13

CHAPTER

A DISASSEMBLY PLANT

Picture in your mind the vehicle assembly plant at the Ford Rouge Center in Dearborn, Michigan. Raw materials, components, and parts flow in – iron ore from Upper Michigan and Minnesota, rubber from Brazil, and countless components from all over the world. Metal, glass, textiles, leathers, plastics, paints, and thousands of other materials go into the making of an automobile. They all funnel into the various parts of the complex – engine, stamping, frame, body, tool and die, paint shop, and assembly plants – and a Ford F-150 pickup truck dives out the other end.

Now run that picture in reverse and you will see metaphorically what goes on at any large meat-packing plant in Colorado, Montana, Kansas, Alberta, or Ontario. Cattle go in one end and hundreds of cuts of beef, thousands of pounds of hamburger, and tens of thousands of non-food products ranging from fertilizers to drugs come out the other end.

Prior to the 1880s, the North American meat industry was small and fragmented. Only the wealthy ate meat on a regular basis. Most people's consumption of meat was restricted to special events when an old cow or sheep was slaughtered and eaten.

Before refrigeration, an animal that had been slaughtered had to be eaten quickly before it spoiled, so it was impossible to distribute the meat widely. As a result, only one family would eat each animal, with perhaps limited distribution to local people.

The industrial revolution produced rapid economic growth by the end of the nineteenth century – a growth fed in large measure by beef. The economic boom resulted in a large segment of the population becoming urbanized and wealthy enough to include meat as a regular part of their diet.[1]

This fact and the advent of refrigeration led to the birth of the modern meat-packing industry. Refrigeration allowed for meat to be stored and shipped, so no longer was it necessary for a slaughtered animal to be consumed quickly.

The demands made on the Union in the American Civil War for foodstuffs and other goods for troops in the field helped to spark the rise of Chicago as a great metropolis, the city immortalized by Carl Sandburg in his poem, "Chicago."

> Hog Butcher for the World,
> Tool Maker, Stacker of Wheat,
> Player with Railroads and the Nation's Freight Handler.

In 1864, a consortium of nine railroad companies began work on what was to be one of the largest industrial structures in the world, the Union Stock Yards in Chicago. They paid $100,000 for 320 acres of swampy land in southwest Chicago. Along with private investors, the railroad companies then provided another $1.6 million to develop the stockyards. By 1900, this complex occupied 475 acres – almost one square mile – and contained within it more than fifty miles of roads, offices, hotels, restaurants, and saloons.

A huge labyrinth of 2,300 livestock pens held up to 200 horses, 21,000 cattle, 22,000 sheep, and 75,000 hogs.[2] Along with the livestock pens were the giant meat-packing plants owned by rivals Armour and Swift. To service the needs of the packing houses, 130 miles of railroad track circled the perimeter of the complex. The yards pumped more than half a million gallons of fresh water every day from the Chicago River, and returned so much manure and waste that the river downstream from the

complex became known as "Bubbly Creek." Later, they drilled two arte-
sian wells, each more than one thousand feet deep, to service the area.[3]

At the turn of the century, the U.S. meat-packing industry employed
68,500 workers (including 3,000 women and 1,700 children).[4] More than
25,000 of them worked in Chicago. At that time Chicago produced 82
per cent of the meat consumed in the United States. Here, millions of
animals were converted into meat and associated by-products.

Thousands of employees spent their days slaughtering and disassem-
bling the animals. A desire to speed up the business of converting pigs
and cows into pork and beef helped to stimulate the birth of the assembly line.

Before the advent of the assembly line, a steer would be taken into the
slaughterhouse, stunned by a hammer blow to the head, then stabbed
and allowed to bleed out onto the floor. Workers then wrestled the carcass onto a frame, where a butcher would cut it up. From the time a cow
entered the slaughterhouse until the butcher began his work took about
fifteen minutes and required three men. This process proved too slow
for the huge numbers of animals that were arriving every day.

Shortly after the union yards were opened, the owners hit upon a rad-
ical new idea for streamlining the operation – the conveyer belt. Now
cattle were stunned in the usual manner, but then a "shackler" would
fasten a chain around the animal's rear legs and hoist it up on a moving
chain powered by a steam engine. One shackler could hoist seventy animals every minute. At the first station, the animal's throat was slit, per-
mitting the blood to flow out as it moved down the line. The task of
butchering was broken down into a multitude of simple repetitive steps.
One person would cut off the head, while the next removed the hooves
and tail. The next might gut the animal and remove all the internal
organs. Skinners removed the hide. As the animal moved along on the
chain it would be quickly and efficiently dissembled.

As the speed of disassembly increased, so labour costs fell. Unskilled
workers needed little training to perform their allotted task. Low-paid
labourers, rather than higher-paid butchers, could butcher the animals.

Henry Ford credited, at least in part, his inspiration for the moving
automobile assembly line to the overhead trolley that had been developed by the meat-packers.[5] (Ransom Eli Olds, founder of the Olds Motor

Works in Detroit, actually built the first stationary automobile assembly line, but Ford improved on the idea by installing a conveyor belt.)

Working in a meat-packing plant must have been as close as anyone has come to hell on earth. Immediately on entry, workers and visitors alike would find all their faculties simultaneously assailed. First was the noise – added to the din of the steam engines were the mechanical sounds of the chain clanking and the bellows of terrified cows who had yet to be stunned or those that had not been properly stunned. At the same time, the smell of cattle and pig excrement and the smell of blood were overwhelming. Because these plants were not air-conditioned, the heat was terrible. Men wielded sharp knives and saws while standing on floors slippery with blood and entrails, all the while trying to keep pace with a line that never stopped.

Regrettably, conditions remain abysmal in most packing plants today. In January 2000, Nebraska Lieutenant Governor David Maurstad reported on working conditions in meat-packing plants in his state. In a model of understatement, he notes that:

> The conditions inherent to meat packing plants create an environment where the work is physically demanding and aesthetically unpleasant. Employees are exposed to significant hazards such as falls, noise, cuts and amputations as well as repetitive motion disorders, back injuries, and other ergonomically related hazards.[6]

It is no wonder that this industry had and continues to have, along with a high labour turnover rate, one of the worst safety records of any industry in the United States.[7]

CONSPIRACY, CONSOLIDATION, AND CONTROL

Almost as soon as the Chicago yards were opened and the assembly-line system initiated, the five companies that controlled the stockyards entered into a conspiracy to control the entire industry. So complete was their hold on the meat-packing industry that U.S. President Woodrow Wilson asked the Federal Trade Commission (FTC) to investigate their affairs.

William Colver chaired the investigation. On July 3, 1918, the committee presented its findings and recommendations in a letter to President
Wilson. This letter denounced the meat-packers unreservedly. Having
been asked whether "monopolies, controls, trusts, combinations, con-
spiracies, or restraints of trade out of harmony with the law, and public
interest" existed, the FTC reported that "we have found conclusive evidence that warrants an unqualified affirmative." The report continued:

> It appears that the five great packing concerns of the country
> – Swift, Armour, Morris, Cudahy, and Wilson – have attained
> such a dominant position that they control at will the market
> in which they sell their products, and hold the fortunes of
> their competitors in their hands.
>
> Not only is the business of gathering, preparing, and sell-
> ing meat products in their control, but an almost countless
> number of by-product industries are similarly dominated;
> and not content with reaching out for mastery as to com-
> modities which substitute for meat and its by-products, they
> have invaded allied industries and even unrelated ones. . . . If
> these five great concerns owned no packing plants and killed
> no cattle and still retained control of the instruments of
> transportation, of marketing and of storage, their position
> would not be less strong than it is.
>
> The producer of livestock is at the mercy of these five
> companies because they control the market and the market-
> ing facilities and to some extent the rolling stock which trans-
> ports the product to the market.
>
> The competitors of these five concerns are at their mercy
> because of the control of the market places, storage facilities,
> and the refrigeration cars for distribution.
>
> The consumer of meat products is at the mercy of these
> five because both producer and competitor are helpless to
> bring relief . . .
>
> Out of the masses of information in our hands, one fact
> stands out with all possible emphasis. The small dominant

group of American meat packers is now international in their
activities, while remaining American in their identity. Blame
which now attaches to them for their practices abroad as well
as at home inevitably will attach to our country if the practices continue.[8]

The committee outlined several reforms as to how this situation could
be remedied.

Over the next eighty years, the U.S. Government has repeatedly
attempted to increase competition in the meat industry. But most of
these attempts have resulted in little more than a slap on the wrist for
the industry.

It is a fact that American companies produce the cheapest food in the
world. (The term *cheapest* here means that Americans pay less of their
annual income for food than almost any country in the world.) And this
fact may partly explain government complaisance. Furthermore, from a
public-health perspective, there are remarkably few poisonings related
to processed food in the United States. The government attitude seems
to endorse the idea that while a few corporations may exercise control
over the food supply, well, they're doing a good job. Furthermore, the
United States produces almost one-quarter of all the beef in the world –
a global giant in beef production, unmatched by any other country.

Currently in the U.S. the largest four companies account for 70 to 80
per cent of the total U.S. market. There has also been a massive consolidation and enlargement of individual cattle slaughter plants. Between
1995 and 2000, the total number of facilities slaughtering cattle declined
by 98 (12 per cent), while the numbers of animals killed at the largest-
volume facilities increased. In 2000, out of a total of 738 cattle slaughter
facilities, 16 plants each killed more than one million animals per year.[9]
This represents an enormous number of animals. If these plants operate
24 hours a day, 365 days a year, each plant will be slaughtering an average
of 2,750 animals a day, 115 an hour, or 2 per minute. (National Beef
slaughters 390 animals every hour, a rate that is not unusual in the
U.S.[10]) These 16 plants slaughter more than 70 per cent of the cattle destined for meat in the entire United States.[11]

In addition, the same four large companies that control the slaughter of cattle also control a very large segment of the grain sector, and thus also control the cattle feed industry.

William Heffernan, professor of rural sociology at the University of Missouri, likens the food system to an hourglass "in which farm commodities produced by thousands of farmers must pass through the narrow part of the glass that is analogous to the few firms that control the processing of the commodities before the food is distributed to millions of people in this and other countries." He adds, "Increasingly, the major decisions in the food system are being made by an ever-declining number of firms. . . . They are primarily concerned with maximizing their profits. That is the purpose of such corporations."[12]

Indeed, the largest of these transnational agri-business firms, Tyson Foods, boasts pro forma revenues nearing $25 billion, with 120,000 "team members" in more than 300 facilities and offices in 32 U.S. states and 22 countries from Canada to Mexico, England, Ireland, China, Japan, Russia, South Korea, and Taiwan. The world's largest processor and marketer of beef, chicken, and pork products, Tyson Foods owns IBP, formerly Iowa Beef Processors, which is the world's largest supplier of beef and pork, as well as allied products, such as tanned hides used to make leather.[13]

Tyson and the other top three U.S. beef packers, ConAgra, Cargill (owner of Excel Corporation), and Farmland National Beef Packing Company, control 81 per cent of the market. Similarly a handful of producers dominate other sectors of the food industry, with many of the same names revealed as the key players across different sectors such as beef and pork packing, poultry, animal feed, grain terminals, corn exports, soybean exports, flour milling, soybean crushing, and ethanol production.

Because grain traders and processors own some of the largest livestock feeding and slaughter operations, we end up with what is called a "closed, vertically integrated market." The Institute for Agriculture and Trade Policy explains:

> In a vertically integrated market, the different stages of production – from corn to the crushing plant to generate animal

feed, high fructose corn syrup and ethanol, to the feeding of cattle on a feeding lot – are internal to a company's operation. There is no price discovery at the different stages of production, meaning no competition to indicate what prices for different operations should be.[14]

But integration doesn't end here. Rendering facilities process by-products such as offal (entrails and internal organs), bone, blood, and other material into fat and protein products for feed and industrial uses. In the past, most rendering facilities operated independently, but the continuing trend is toward increasing volumes processed by packer/renderers. Because many of the largest meat packers (especially in poultry) are integrated right from feed manufacturing and livestock production all the way through slaughter and distribution, the animal proteins produced in their own rendering facilities are sometimes used in their own feed rations.

What does all this mean to our ability to control prion diseases? This great concentration and integration of industry may serve to magnify the chance of widespread distribution of a relatively rare event. Let us suppose that a single prion-affected cow enters a packing plant with no obvious symptoms of disease. Material from this cow could go not only into beef for human consumption (conceivably being incorporated into thousands of pounds of ground meat for hamburgers) but also to the rendering plant attached to the slaughterhouse and to the feed made with material from the rendering plant. Of course, as a we learn more about the dangers inherent in prion diseases, the concentration and integration of the industry may also make it possible to implement safeguards quickly and efficently.

FATTENED FOR SLAUGHTER:
WITH FEATHERS, GARBAGE, AND MANURE

A newborn calf weighs about 80 pounds. By the time it goes to slaughter, the average cow weighs about 1,213 pounds (up from 1,187 pounds in 1995). What has changed over the last hundred years is the length of time it takes for a calf to put on that additional 1,133 pounds. The methods of feeding and fattening cattle have undergone a complete transformation. Back when cattle roamed the vast pasturelands of North America, they pretty much kept their own schedule of feeding and fattening. No longer. A fourth-generation cattle rancher quoted in the *New York Times* says, "In my grandfather's day, steers were 4 or 5 years old at slaughter. In the 50's when my father was ranching, it was 2 or 3. Now we get there at 14 to 16 months."[15]

Beef processors have almost universally adopted the feedlot system of "finishing" cattle. Finishing refers to the large weight gain that cows undergo between the ages of about eight months and fifteen months. This time is spent in huge penned enclosures with hundreds or thousands of other cattle being fed a scientifically formulated food designed to promote the maximum weight gain in the shortest possible time. Once in the finishing process, cows gain about three and a half pounds every day. Not only is this an efficient and economic method of feeding cattle, it also produces a uniform and somewhat fatty beef.

Because meat is mainly protein, specialists in animal nutrition recognized very early that one of the keys to rapid weight gain was the amount of protein in the animal's diet. In 1939, Roscoe Snapp, professor of animal science at the University of Illinois, noted the "need of the animal body for the complex organic compounds called proteins. The important thing about proteins is that they contain nitrogen, an element indispensable to all animal life. . . . As a matter of fact, the lean portions of the body, the skin and its modifications, and the connective tissues consist almost entirely of protein materials, which have been elaborately built up from the protein stuffs consumed by the animal."[16]

Unfortunately, corn delivers only about 9 per cent protein while grasses such as alfalfa contain about 18 per cent protein. Other vegetable-derived foodstuffs such as hay, or various grain stuffs that have traditionally made up the bulk of cattle feed, all are in the range of 10 to 20 per cent protein. From the earliest days of feeding cattle the search was on for an inexpensive high-protein feed, or at least a food supplement.

U.S. farmers began forcing cows to become cannibals in the early 1940s. A 1947 handbook on the feeding of beef cattle instructed farmers that meat scraps and products from rendering plants "are satisfactory protein supplements for beef cattle."

> Sometimes cattle may not like meat scraps or tankage at first, but after several days they will generally eat the small amount that is needed to balance a ration. It is best to accustom a cow to these feeds gradually by mixing a small proportion at the start with better-liked feeds.[17]

Perhaps the cows were trying to tell their feeders something.

In the search for increased usable protein to be included in feed, no substance was off limits, including "recycled" cow manure and chicken manure as food for cows. In 1972, governments expressed concern about the potential environmental hazard presented by the huge quantities of manure from feedlots. At that time, U.S. farm animals produced an estimated 112 million tons of manure dry matter per year (obviously much heavier when fresh), while Canadian animals produced 24 million tons. These two problems – increasing protein in feeds and getting rid of manure – taken together suggested a brilliant solution.

Animal nutritionists suggested that animals, especially ruminants, could be placed in "an improved competitive position for supplying an even greater part of the world's future protein needs." (A ruminant animal has four stomachs; ruminants include cattle, sheep, goats, deer, elk, and bison. This proposal targeted cattle.) To that end, they promoted "animal recycle systems for utilizing undigested feed nutrients and by-products of digestion and metabolism." In plain English, it would make great economic sense if, rather than spreading cow manure on grass and then using the grass for feed, we can skip the middle step, i.e., grass, and feed the manure directly to the cattle. If they could find a way of doing this cheaply, they would have a winner on their hands.[18]

If this proposal makes you feel queasy, you have only to look at the ingredients currently used in formulating cattle feed. Among the more than 175 items, we find dried cattle manure with a protein content of 17 per cent, blood meal, hydrolyzed feather meal, hydrolyzed animal hair, cooked municipal garbage, ground limestone, dried poultry manure, and numerous chemicals such as urea, sodium triolyphos, and monoammonium phosphate. Of all of these, blood meal with 80 per cent protein is one of the most desirable, followed by meat meal at 55 per cent.[19] We know where manure comes from, but what is the source for feather meal, or blood and meat meal? These are products of a huge – and very low-profile – sector, the rendering industry.

CHAPTER 14

EVERYTHING BUT THE MOO

After a cow is killed, it is taken apart. What we consider meat – steaks, roasts, and chops – are cut up and packaged. Pieces of meat trimmed off in this first process are collected, mixed with fat, and made into ground beef. But a 1,213-pound cow doesn't produce 1,213 pounds of meat. Hides account for about 5 per cent of the weight, while other by-products make up 34.1 per cent. This means that a live slaughter weight of 1,213 pounds produces two sides of beef with a total weight of under 800 pounds.

David Lamb, former operator of a small-town custom butchering business in Saskatchewan in the early 1970s, remembers that modest operations differed greatly from large meat-packing plants.

> In the small plants there was much more waste than in the large plants. The heads and a lot of the bones (brains and spinal column included) were just thrown out since there was no way to process them. The big plants render and grind all of these by-products.[1]

Hides are removed intact and sold to tanning companies to be made into leather. The meat enters the food distribution system. Everything else that comes out of a meat-packing plant goes to the renderers. This

includes animal offal, which is made up of viscera (intestines and inedible organs), heads, bones, blood, and other waste products. Rendering
facilities (usually independents) also process any animal that dies before
it reaches the slaughterhouse, as well as dead pets, horses, and waste
from restaurants and butcher shops. However, as rendering fees increase,
more farmers tend to bury their dead livestock on the farm.

Three major renderers operate twenty-six plants in Canada, out of a
total of about thirty-five renderers in the industry. Every week these renderers process thirty thousand tonnes of animal wastes and by-products,
which amounts to nearly 3.5 billion pounds per year.[2]

The United States slaughters 139 million head of cattle, calves, sheep,
hogs, and horses, and 36 billion pounds of poultry a year. This enterprise
produces a staggering volume of animal parts not fit for human consumption – animal parts that go to rendering plants. The U.S. produces
more than 47 billion pounds of material to be rendered every year, a
volume handled by just 260 renderers. It has been said that if this material were placed in trucks, the procession of trucks would stretch across
the U.S. from New York City to Los Angeles four times.

Rendering plants turn this enormous mass of otherwise unwanted
waste into useful and profitable commercial products. Right from the
early days, the packing houses built factories to deal with the slaughterhouse by-products. These on-site or nearby factories produced leather,
soap, fertilizer, glue, imitation ivory, gelatin, shoe polish, buttons, per-
fume, and violin strings.

By-products account for more than 10 per cent of the value of a steer,
and keep beef prices to consumers lower than they otherwise would be.
Today, by-products include:

- Hides, which become leather for shoes, upholstery,
  luggage, garments, footballs, and so on. Hides are the
  single most valuable by-product.
- Tallow, which is used in animal feed and soap, as well as to
  make glycerin and various acids (oleic, stearic, linoleic).
  These by-products are used in everything from cosmetics,
  creams, shampoos, perfumes, and toothpaste, to candles,
  crayons, rubber tires, paint, glue, medicines, pill capsules,

solvents, textiles, and explosives.³ Vegetarians and vegans
who hope to avoid animal products face a vast array of
goods in the marketplace that contain numerous hidden
sources of animal by-products.

- Meat and bone meal (MBM) and blood meal, the most
  commercially important by-products by far – valued
  for their low cost relative to their high protein content.
  In 2000, U.S. renderers produced almost seven billion
  pounds of MBM and blood meal. Almost all of this
  MBM and blood meal is sold to livestock and poultry
  operations, as well as to pet-food manufacturers for
  inclusion in feed rations. Some is used in fertilizers.

These by-products infiltrate every aspect of our everyday lives, as we
shall see.

In keeping with the trend toward consolidation in the industry, many
rendering plants are integrated into the giant slaughterhouse plants. In
1995, the production of MBM in the U.S. was roughly split between independent renderers and those associated with large slaughter facilities. By
2000, the packer/renderer operations were producing at least 60 per cent
of the MBM nationally. The concentration of ownership also extends
beyond the packer/renderer operations. The same companies that own
the slaughtering and rendering plants now own a large and growing
number of feed manufacturers.

DOWN BUT NOT OUT

Animals that can't stand up – so-called "downer" animals – provide
another important source of material for rendering plants. These animals may be non-ambulatory and unable to rise because they have a
metabolic disease, broken limbs, or suffer from a central nervous system
disorder.

Numerous studies from European countries have shown that downer
cattle are much more liable to have mad cow disease than the general
population. A 2001 study from Germany found that downer cows were
anywhere from 10 to 240 times more likely to test positive for BSE than

were ambulatory cows. A USDA study concluded that if mad cow disease ever did occur in the United States, it would most likely be found among downer cattle rather than randomly in the general cattle population.

In fact, that is exactly what happened in 2003. Two cows tested positive for BSE, one in Canada, one in the U.S. Both were downer cows.

Agriculture Canada's code of practice suggests that sick, injured, or disabled cattle in severe distress should be euthanized or slaughtered on the farm, not transported to slaughterhouses. But no laws back up this recommendation. So downer cows can and do get slaughtered and approved for human consumption.

A Canadian Food Inspection Agency directive notes that severely crippled or downer animals may be stunned and bled in public stockyards before transportation to a registered slaughter establishment under certain conditions.

But a telling endnote should ring a warning bell:

> It should be noted that meat products which are approved for human consumption from such carcasses are not eligible for export to any country.[4]

So even though nobody else will take this meat, Canadians are free to eat it! And even though the single mad cow discovered in Canada did not get into the human food chain, there's no way of telling whether undiscovered cases may have done so.

Many U.S. cattle slaughterhouses enforce strict policies against processing downers for human consumption, but U.S. regulations did not prohibit the practice until December 30, 2003, after the discovery of BSE in a downer cow in Washington state. Prior to that time, most downer animals brought to USDA slaughterhouses were approved for human food.

In a curious contradiction of principles, even though they continued to allow meat from downer animals to be sold for human consumption, last year the USDA announced a landmark policy to stop purchasing meat from downer cows for federal programs, including the National School Lunch Program. Many big fast-food chains and larger supermarkets have recognized the dangers of meat from downer cows – McDonald's, Wendy's, and Burger King, along with supermarkets such as Safeway and Albertsons, all refuse to accept the meat of downer animals.

But until this year meat from downer cows was still sold in many supermarkets, restaurants, and butcher shops. Now the only remaining market for downer cow meat in the U.S. is pet foods.

In Canada, downer cattle are *usually* turned into pet food, but if they can be kept alive long enough, they can still be used directly for human food. At this writing, Canada has not yet taken the obvious precaution adopted by the U.S. in banning such usage. Only a "sample" of downer cows determined to be safe for human consumption are tested for BSE and held pending negative test results.

If that makes you uneasy, the news only gets worse. Meat from "dead-stock" still makes it to the retail level, and into consumers' kitchens. "Deadstock" are animals that die either in the field, or in transit, often due to illness, before they reach the slaughterhouse. Ontario, like most jurisdictions in Canada and elsewhere, prohibits the processing or sale of meat from dead animals for human consumption. Deadstock is supposed to be used for animal feed only. Inspectors are supposed to examine animals immediately before slaughter. The very sound theory behind this prohibition is the need to observe an animal's behaviour while it is still alive in order to detect neurological or other problems.

Yet in August 2003, as part of an undercover investigation, Ontario provincial investigators witnessed the butchering of dead animals at Aylmer Meat Packers (the owner of which had a history of harassing inspectors, and a conviction for assaulting a veterinarian). At least some of the deadstock processing is alleged to have happened after-hours, when inspectors were not present. At the time of writing, Aylmer Meat Packers, which denies the allegations, is the subject of a criminal investigation.

Testing Downer Cattle

At a typical North American feedlot operation, downer cattle compose slightly more than 1 per cent of all cattle in the feedlot. A similar number die for no obvious reason on dairy or beef farms every year. The World

Health Organization recommends that all downer cattle be tested for BSE. Japan and most European countries test 100 per cent of all adult downer cattle for the disease.

Many countries are not content to examine only downer cattle. For years Germany insisted that it had no BSE. But when independent testing companies were invited into the country – against the wishes of the government – Germany suddenly discovered that it did indeed have a significant mad cow disease problem. Germany now tests one out of every three cows going to market. Japan and Ireland test every cow they slaughter, not just downer cows, before releasing meat for sale.

By contrast, over the past decade, Canada and the United States have tested only a tiny percentage of downer cattle, although both countries boast of their aggressive BSE surveillance systems and of testing that exceeds the internationally recognized standard for appropriate surveillance. Canada's beef cattle herd numbers about 11 million. Surveillance has increased greatly since the infamous mad cow of 2003, but many believe that the sample tested is still far too small.

In 1993, the U.S. mad cow surveillance program expanded to include examination of brain tissue from downer cows. The U.S. is quick to point out that the tissue from only one cow has revealed evidence of BSE and that cow came from Canada. But some say this amounts to a "don't look, don't find" policy.

From 1993 to 2001, there had to be at least 3 million downer cattle processed from U.S. feedlots alone (1 per cent multiplied by 37 million cattle slaughtered per year multiplied by nine years). This number does not include animals such as farm dairy cattle, pigs, and sheep that died in this period that were also sent to the rendering plants. The 20,141 cattle whose brains were examined during that period represent far less than 1 per cent of those that were rendered.

Dr. Stanley Prusiner says he would like not only every downer cow tested but eventually would like to see *every* cow tested for BSE. On a 2003 National Public radio broadcast, he told listeners:

> We have about a hundred million cattle in the United States, and we slaughter about a third of these every year. About one percent of this hundred million cattle are called downer

cattle. . . . And these animals just disappear. And when we talk
about 19,000 cattle [tested in 2002 in the United States for
mad cow disease] – looking at those animals . . . most of that
number represents fallen cattle. But that's 19,000 out of a million. So I think those numbers [of cattle tested in the U.S.]
are appalling.[5]

Until Canada diagnosed its single case of BSE in 2003, its cattle industry was closely integrated with that of the United States. In the preced-
ing year, Alberta shipped more than half a million live cattle to the U.S.
The two countries also traded large volumes of meat and bone meal for
feeding to animals. That practice stopped cold when Canada announced
it had found a case of BSE in a cow in Alberta – the heartland of Canada's
cattle country.

When the second mad cow born in Alberta turned up in Washington
state late in 2003, the ministers of agriculture of Canada, the U.S., and
Mexico agreed to begin development of global incentives to further the
control and eradication of BSE. In 2004, Canada announced an increase
in BSE testing levels, with at least 8,000 animals to be tested in the first
year, rising to testing levels of 30,000 animals a year or more.

Rendering and Risks

Regardless of the cause of death of downer cattle, the USDA says that 95
per cent of them are sent to the renderers.[6] (In Canada, less than 10 per
cent of raw material for rendering comes from deadstock; most comes
from meat packers.[7] Some renderers do not accept dead pets or roadkill.)

Given that most of the downer cattle are rendered into animal feed,
should we just accept the risk that some of them become meat for
human consumption in Canada and the smaller risk that some of them
may harbour BSE?

Remember that meat and bone meal (MBM) from rendering plants is
sold to livestock and poultry operations as well as to pet-food manufacturers for inclusion in feed rations.

The long and extensive inquiry that resulted from the mad cow (BSE)
epidemic in Britain reached a number of key conclusions. One main

conclusion was that there was a link between the manufacture of meat and bone meal (MBM) for cattle feed and the spread of BSE.

How MBM is made has undergone several evolutions in the past one hundred years. The original process consisted of the mechanical removal of any meat or other tissue from the carcass and then grinding up what remained. Gradually the process became more mechanized. Cutting machines such as drills or augers were added to ream out any meat or fat inside the bones. High-pressure water jets flushed out anything the cutting machines missed. Finally, chemical solvents were used to extract the last bit of protein that remained. The result was a pinkish slurry that was dried and made into pellets mainly for the animal food and fertilizer industries.

The BSE Inquiry determined that MBM in cattle feed was the vector responsible for the BSE epidemic. Thus, BSE arose as a result of cow cannibalism – just as kuru had arisen from human cannibalism. But if cows eating infected cows caused more cows to be infected, where had the original infection come from? Today, opinion remains divided in the scientific community.

- One line of reasoning holds that the probable original source of infection was scrapie-contaminated MBM. Scrapie-infected sheep must, however, have been a constituent of the MBM that went into cattle feed for decades. Why had it only started to transmit to cattle in about 1982? Those who favour this argument suggest that MBM had become infectious because rendering methods, which had previously inactivated the conventional scrapie agent, had been changed. The BSE Inquiry rejected this theory, finding it unlikely that there had been a change, or a combination of changes, in rendering that was peculiar to Great Britain and yet so nearly universal within Great Britain that they permitted an almost simultaneous transmission of scrapie to cattle throughout the country.
- Another line of inquiry suggested that the cause of BSE might be manganese and pesticides. In the early 1980s, the

British government mandated the systemic use of
organophosphate insecticides to combat warble-fly
infestations. These nerve poisons might have combined
with high-manganese mineral supplements to disrupt
normal brain functioning and cause the mad cow
outbreak. The BSE Inquiry concluded that this theory
was not viable, although there is a possibility that the
manganese and pesticides could increase the susceptibility
of cattle to BSE.

- The BSE Inquiry ultimately concluded that the most likely
possibility is that BSE is a novel and unique TSE that
probably started as a consequence of a genetic mutation
in a single animal, and which was propagated as a
consequence of recycling by rendering and incorporation
in cattle feed. This would constitute a point source for the
epidemic. Their ultimate verdict: "The origin of the
disease will probably never be known with certainty."[8]

In Britain, when authorities finally admitted that infected cattle feed
had been responsible for the mad cow epidemic, they first banned the
feeding of any cow MBM to cows. Later they banned the feeding of sheep
MBM to cows. Eventually the European Union (EU) banned the feeding of
all MBM to all farmed animals. The ban, which was to have been temporary, has been extended several times because failure to enforce the regulations meant that the member states could not ensure that ruminant
by-products such as MBM did not contaminate other animal by-products.

The ban, of course, has led to the huge problem of what to do with all
those enormous stockpiles of rendered animal parts – some of them
potentially BSE-contaminated – now that they can neither feed them
back to cattle or other animals nor export them. By the end of 2000, a
mountain of 3 million tonnes of animal meal had built up, including 2.5
million tonnes of MBM. The EU is incinerating the MBM to dispose of it,
but with a backlog of more than a million tonnes still waiting for disposal, there are reports that potentially BSE-contaminated material is
now being mixed with clean material.[9]

Other jurisdictions around the world, notably Japan and Eastern Europe, have followed the EU's lead in banning MBM in feed for all animals, a fact that has angered U.S. producers of MBM, as they have seen their markets for the feed dry up.

In 1998, Vermont learned that three flocks of sheep may have been exposed to BSE-contaminated feed in Belgium and the Netherlands, from where they originated. All three flocks were quarantined. In July 2000, four of those sheep in Vermont tested positive for TSE. In December, the U.S. government banned all imports of rendered animal protein products from Europe, regardless of species.

We already know that human beings developed vCJD after eating BSE-infected meat in Britain. In the United States, mink that were fed a diet including fresh meat products from downer cattle developed transmissible mink encephalopathy (TME).[10] Investigators theorized that these downer cattle introduced a TSE agent (a prion) to the mink, which caused the mink to develop the brain-wasting disease. On autopsy, the mink brains showed spongelike holes in the brain – holes similar to those found in the brains of CJD victims and other TSEs. If there really were no BSE in the United States, what caused those mink to develop TME?

Here's where the food chain gets a bit complicated. Canada and the United States prohibit feeding of rendered ruminant (cow, goat, sheep) protein (MBM) to ruminants, but they do allow *any* rendered protein to be fed to *non-ruminants*. So MBM made from downer cows may be fed to pigs, horses, and chickens. In fact, the Canadian Food Inspection Agency (CFIA) admits that feed banned for cows is routinely fed to pigs and poultry, since there is no evidence (according to the CFIA) that those animals can contract BSE.

But in turn, approved animal protein products – porcine meal, horse protein, and poultry – may be fed to cows. In other words, you can feed your pigs and chickens with infected MBM – then you can take those same pigs and chickens and make them into MBM to feed back to your cows.

(Cattle may also be fed milk, blood, and gelatin [made from rendered cows] and non-protein animal products, such as rendered animal fats, e.g., beef tallow, lard, poultry fat.)

Now the chain looks something like this:

Downer cow (possibly BSE infected) –> rendered into MBM
–> fed to pigs and chickens –> rendered into MBM –> fed to
cow –> human consumption.

Or it can also look like this:

Downer cow (possibly BSE infected) –> rendered into MBM
–> fed to pigs and chickens –> human consumption.

European experience shows that accidental or intentional mixing of
feeds either at the feed mill or on the farm provides an even more direct
route of contamination. Farmers have often used feed intended for one
species to feed another should they run short.[11] There's no guarantee
that the bag of prohibited feed intended for the pigs won't go to the
cattle on a mixed farm.

This fact has not escaped the notice of the USDA. In January 2003, it
solicited public comment "to help . . . control the risk that dead stock
and nonambulatory animals could serve as potential pathways for the
spread of bovine spongiform encephalopathy, if that disease should ever
be introduced into the United States."[12]

Only thirteen Canadian rendering plants produce both prohibited
and non-prohibited material – all with "processes in place to reduce the
risk of cross-contamination."

High-producing dairy cattle are more likely than beef cattle to receive
protein supplements containing MBM. These same dairy cattle are the
ones likely to be turned into hamburger at the end of their lives. In view
of the mad cow debacle – and the $30-billion industry across the conti-
nent – the CFIA is contemplating whether Canada should consider a
total ban on all MBM to *all* farmed animals, as has been done in Europe.

WHISTLING IN THE DARK

Whether authorities are being disingenuous or just whistling in the dark
is perhaps a moot point. It's well documented that prion diseases (TSEs)
exist in numerous species in the United States and Canada, including

sheep, elk, deer, and mink. Some of these infected animals were certainly rendered into cattle feed. Both the U.S. and Canada fed MBM (meat and bone meal, including rendered cattle and sheep) to cattle until just a few years ago.

In 1997, both Canada and the U.S. banned the feeding of all rendered ruminant proteins to ruminants. In Canada, the CFIA assures us that a "high level of compliance" has been recorded among feed manufacturers. If non-compliance is noted, CFIA inspectors secure corrective action and confirm this with follow-up inspections.

But according to many, the practice continues both illegally and inadvertently, in both Canada and the U.S. After some cattle in Texas accidentally received ruminant MBM feed, several U.S. feed companies stopped using ruminant MBM in *all* feeds in order to prevent contamination of cattle feed.

Inspection of renderers, feed mills (licensed and unlicensed), ruminant feeders, on-farm mixers, protein blenders, and distributors in the U.S. has shown different levels of compliance with this feed-ban regulation. In 2001 and 2002, the FDA found hundreds of firms out of compliance. Non-compliance consisted of one or more offences such as:

- Products not labelled as required (e.g., feed containing proteins derived from mammalian tissue such as meat and bone meal derived from cattle must be labelled with the caution statement "Do Not Feed to Cattle or Other Ruminants").
- Inadequate systems to prevent comingling of feed for cattle with prohibited feed.
- Inadequate record-keeping.

When the FDA reinspected firms that were out of compliance, they found hundreds that still had not changed their ways.

In one instance in 2003, the FDA filed a Consent Decree of Permanent Injunction against X-Cel Feeds Inc., headquartered in Tacoma, Washington, based on violations of the Food, Drug, and Cosmetic Act. In the decree, the firm and its officers admitted liability for introducing adulterated and misbranded animal feeds into interstate commerce and agreed to implement measures to correct the violations under the FDA's supervision.[13]

RISKY MATERIAL

Paul Brown, senior research scientist at the Laboratory of Central Nervous System Diseases at the U.S. National Institutes of Health, notes that the major suspect for the contamination of beef products is mechanically recovered meat (MRM). This is a paste made from compressed carcasses from which all other consumable tissues have been manually removed. In December 1995, new regulations prohibited certain materials such as spinal cords and paraspinal ganglia from inclusion in mechanically recovered meat in the United Kingdom. But up until that time, these items from all slaughtered animals were routinely included in MRM. This product was *legally defined as meat* and was permitted to be included in most cooked meat products, such as hot dogs, sausages, meat pies, tinned meats, luncheon meats, and precooked meat patties.[14]

In BSE-infected cattle, the infective agent is concentrated in tissues such as the brain and spinal cord. In July 2003, Canada introduced regulations to prevent specified risk material (SRM) from entering the human food supply. SRM are defined as the skull, brain, trigeminal ganglia (nerves attached to the brain), eyes, tonsils, spinal cord, and dorsal root ganglia (nerves attached to the spinal cord) of cattle aged thirty months or older, and the distal ileum (portion of the small intestine) of cattle of all ages.

These regulations require that all SRM are to be diverted away from the food supply. Although the issue had long been "under consideration," the United States did not declare an SRM ban until December 30, 2003.

PART FIVE

A Modern Plague

CHAPTER 15

SMALLPOX, SYPHILIS, AIDS . . .
AND ALZHEIMER'S?

We use several different terms to describe widespread diseases. The word *epidemic* derives from the Greek *epidemios*, meaning "among the people" – especially the common people or citizenry. Hippocrates used the word to describe diseases that were common at a given point in time among the general populace. (As a contrast, the word *endemic* refers to diseases usually found among a particular people or in a certain region – often as a result of permanent local factors.)

The word *pandemic* comes from the Greek *pandemos* – as pan-demos, or all people, and refers to a disease prevalent over a whole country or the world.

The word *plague* comes from both Greek and Latin. Both the Greek *plege* and the Latin *plaga* mean "a blow or a stroke." The Latin *plangere* means "to beat or to strike" (especially one's own breast). It can also mean "to bewail or lament." The meaning implied either a divine stroke for some mortal transgression, or the wailing and lamentation that results from such a blow. The word was originally applied to any epidemic that was particularly destructive resulting in large loss of life.[1]

With the baby boomers fast approaching their sixth decade, we will experience far-reaching consequences – in society, government, health care, medicine, business, and industry – in the next twenty years due to the growing numbers of people with Alzheimer's disease (AD). Already epidemic in proportions, AD will be the plague of the twenty-first century.

THE GREAT PLAGUES

To appreciate the extent of the current Alzheimer's problem, we can compare it to some of the great plagues of history. We'll also see what factors are responsible for the emergence of a plague, and how these factors relate to AD.

Black Death

A pandemic of bubonic and pneumonic plagues swept through Europe from Asia in the mid-fourteenth century. Victims of the Black Death suffer high fevers and aching limbs, followed by vomiting of blood. Lymph nodes in the neck, armpits, and groin swell and protrude, appearing grotesque and black (hence the name Black Death).

The swellings eventually burst, and the victim dies soon after. The whole process – from first symptoms of fever and aches to final expiration – lasts only three or four days. The swiftness of the disease, the terrible pain, the grotesque appearance of the victims, all served to make the plague especially terrifying.[2]

Even today, most children can recite the nursery rhyme "Ring Around the Rosy," but few, if any, would associate it with its origin in the plague. The initially rose-coloured lesions probably gave rise to the rhyme.

- "Ring around the rosy" refers to the rose-coloured area of skin with a darker circle surrounding it.
- "Pocket full of posies" were the sweet-smelling flowers that those tending the sick would carry to ward off the stench of disease.
- "Ashes, ashes" meant impending death (some children say, "A-choo, a-choo," which refers to the sneezing and coughing of pneumonic plague).
- "All fall down" means death. Pneumonic plague has 100 per cent death rate if not treated within the first

twenty-four hours. Bubonic plague has a 1 to 15 per cent
death rate in treated cases and a 40 to 60 per cent death
rate if not treated. Treatment today consists of antibiotics,
which of course did not exist during the great plague of
Black Death.

Infected fleas that live on rats and other rodents spread bubonic
plague. In Venice, according to the city's excellent records, 60 per cent of
the population died over the course of 18 months, 500 to 600 a day at the
height.[3] At the peak of the epidemic, Paris lost 800 people a day. By 1349,
half its population of 100,000 people had died.[4] The pandemic eventually killed 25 million people in Europe – about a third of the population
of the continent, as well as 13 million Chinese.

The Plague with a capital P still refers to the disease caused by the
bacteria *Pasteurella pestis* (also called *Bacillus pestis* and *Yersinia pestis*).
The word with a small p indicates the generic term, meaning a devastating epidemic, usually with a high loss of life.

Resurfacing several times over the next century, the Plague pushed
forward the rise of hospitals, which helped to control the spread of disease. Hospitals played many roles in medieval society – caring for the
poor, vagrants, lepers, the old, and pilgrims, effectively isolating sick
people to protect the well. Affluent citizens could afford to be treated at
home; thus hospitals arguably served as an instrument of social control,[5]
although the Black Death did not play favourites according to class.

Why did the Black Death cause so much death and disease worldwide
but especially in Europe?

First, the germ that caused this disease, *Pasteurella pestis*, was a new
pathogen. The disease had originated in Asia, and the European populace had never been exposed to it before, so there was no level of immunity in the general public.

Many times in history we have seen the devastation that a foreign
germ can cause in a new population. During the "conquest" of North
and South America by the Spanish, British, and French, diseases such as
measles, which were considered minor infections to the European con-
querors, killed hundreds of times more indigenous people than died by
warfare. To the natives of the New World, measles was a new disease, and

deadly to those who were exposed to it for the first time. The same phenomenon can be seen in recent times with the SARS (severe acute respiratory syndrome) outbreak. Once again a new virus, to which people have no native immunity, is proving to be extremely contagious and deadly.

Another reason that the Black Death affected huge numbers of people was that no one knew how the disease was spread. Doctors realized that disease was passed from person to person. During the height of the Plague, homes of Plague victims were nailed shut so that all inside were in effect quarantined from the other citizens of the town. Unfortunately, these early public-health measures had no effect against the primary cause of the spread of the disease: rat-borne fleas. As rats and their fleas dwelt side by side with humans everywhere at that time, confining people with the disease did very little to stem the contagion. The Great Fire of London in 1666, although tragic in its immediate effects, was perhaps one of the most effective public-health measures ever, because it destroyed the squalid residences where the rats thrived. (Only the introduction of sanitary sewage disposal in the late eighteenth and early nineteenth century had an equivalent positive effect on public health.)

This earliest of the great plagues illustrates two of the factors that cause a plague to become established in a society.

- First, a new pathogen (i.e., a germ that has not been endemic to that society) attacks the populace. This means that the disease will be much more virulent than a native disease.
- Second, ignorance as to the method of spread of the pathogen leads to lack of control. This means that effective strategies to control the spread of the disease cannot be implemented.

The Plague continues to plague us – in the United States, an average of eighteen cases per year (mostly very mild illness) have been reported during the last few decades. Outside the United States, the World Health Organization has reported an annual average of 1,666 cases from

1967–1993. The number of actual cases is probably much higher because many countries fail to diagnose and report the Plague.[6]

Other factors can cause a disease to become a plague. The story of the next great worldwide plague, syphilis, illustrates some of these factors.

Syphilis

Until AIDS appeared, syphilis, with a long and nasty history, rated as the worst of the sexually transmitted diseases. With more than 70 million cases worldwide, it still rates as a disease of epidemic proportions. Even though it is now easily cured with penicillin, syphilis remains a major infectious disease. In the past twenty years, there have been 1 million new cases a year of syphilis.

Syphilis victims typically suffer three stages of symptoms, beginning with a genital sore called a chancre that appears between ten days and ten weeks following sexual contact with an infected individual. This chancre is usually fairly small and painless, so in women especially it may go unnoticed. After a week or so the sore heals.

Several weeks to months after the sore has healed the patient enters the secondary stage. The secondary stage usually brings a mild generalized rash. This rash is also painless and often very faint, and it too vanishes within a few days.

Once the rash clears the patient enters the latency period. During this period, which can be as short as two years and as long as thirty years, the patient has absolutely no symptoms, and even if he were aware of the first two stages would consider himself cured.

After the latency period the disease comes back with a vengeance. The third stage of syphilis is characterized by major symptoms, the most common of which is dementia. Not only the brain, but the liver, bones, skin, and great blood vessels can also be affected. Once a person starts to show signs of this late-stage, or tertiary, syphilis, treatment with antibiotics can still arrest the disease. If this is not done, the disease follows a rapid downhill course ending in death.

One well-supported theory holds that Christopher Columbus's crewmembers carried syphilis from the New World back to Italy in 1493.

Charles VIII's soldiers probably took it home to France from their Italian campaign in 1495. Soon the French city of Lyons began to banish infected syphilitics from their city. Within two years, most of France was infected. Less than ten years later, syphilis had spread all over Europe.

Without exception citizens of each country blamed citizens of another for the Great Pox. Muscovites called it the Polish sickness; Poles named it the German sickness. Germans, English, and Italians showed remarkable solidarity in referring to it as the French disease (*morbus Gallicus*). Other names for it included the Disease of Naples, the Italian pox, the Turkish disease, and the Spanish disease.

Because syphilis can be cured easily today, many believe that the disease may be a good candidate for total elimination (as has occurred with smallpox), at least in the United States, if not in the entire world.[7]

Why did syphilis emerge as a plague? Like the Black Death, the disease was introduced into a new population (Europe) that had not previously been exposed to it and so had no native resistance. Early reports of the disease from sixteenth-century France and Holland indicate that the disease was far more deadly then, when it was first introduced, than it is today. Because we have lived with syphilis now for hundreds of years, we have some resistance to it. This does not mean that it is not as infectious, only that cases today are much milder than they were four hundred years ago.

But the main reason syphilis was so deadly was its long latency period. Many years passed between the time a person became infected and the time he or she manifested the severe symptoms of the disease. This had two consequences. First, it took a long time before people were able to make the connection between a sexual encounter twenty years ago and a severe dementia today. Second, because there was this extremely long latency period, many cases of the disease could be present in a society before anyone realized how severe the disease really was. This meant that even if appropriate public-health measures were taken, there still existed a huge reservoir of people in the society who were carrying the disease, and who were unaware of the fact that they would some day come down with the disease.

Syphilis illustrates the third factor that can lead to a new disease becoming a plague:

- A long latency period allows large segments of a population
  to be infected with the disease before overt symptoms of
  the disease manifest.

Smallpox

The variola virus was first documented in 1122, but did not get its
common name, smallpox, until an epidemic in England in 1666.
Smallpox, with its fever, rash, and pustular skin lesions, was called
"small" because the Great Pox was syphilis. The smallpox virus afflicted
citizens of Africa, China, India, and Europe. In eighteenth-century
Europe, smallpox killed as many as four hundred thousand people each
year, and caused one-third of the blindness in the continent. Because of
its high mortality rate, smallpox contributed to the slow rate of increase
in population.

If the New World gave the Old World syphilis, then the Old World
returned the favour in the form of smallpox. The English spread the
ghastly killer when they colonized the Americas in the eighteenth century. The aboriginal people of the Americas had no immunity against it
and were decimated.

Initial symptoms of smallpox resemble the flu but are followed by a
rash, which then evolves into pus-filled blisters. If the patient survives,
these blisters almost always result in severe scarring. Smallpox caused
more than two million deaths per year during major outbreaks. It killed
one in three of those who caught it, and left any who survived it badly
disfigured.

In 1721, Lady Mary Wortly Montague brought home with her to
England a treatment for smallpox from Turkey. This treatment involved
inoculating healthy people with matter from the pustules of those who
had a mild case of the disease. Sometimes it worked – as it did with Lady
Mary – but sometimes the unfortunate recipient of the inoculate died.

The fact that smallpox ceased to be a threat to world health was completely due to the work of an English country doctor, Edward Jenner. In
the eighteenth century, milkmaids often suffered from the mild disease
cowpox, which they acquired from handling cows every day. Jenner

noticed that the milkmaids often refused the smallpox inoculation, claiming they were already immune to the disease. In 1798, Jenner proved them correct when he developed a vaccine from cowpox that completely protected an individual from contracting smallpox. This discovery was all the more important as there is still no treatment for this terrible affliction. Jenner coined the word *vaccination* (from the Latin *vacca*, meaning a cow) for his treatment, a word that Pasteur later adopted to mean immunization against any disease.

Due to mass vaccination, the last case of smallpox in the United States occurred in 1949. But as recently as 1967, some 10 million to 15 million cases of smallpox were still occurring annually in more than thirty countries. At that time, the World Health Organization (WHO) launched a program of mass vaccination of susceptible persons in endemic countries. The most recent outbreak of smallpox in 1977 in Somalia compelled the WHO to lead a collaborative global vaccination program to conquer the disease.

The last recorded death by smallpox occurred in 1978. Janet Parker, a British medical photographer, had been taking pictures in a lab in the Birmingham University Medical School Laboratories when she was accidentally infected with a sample of the smallpox virus stored at the lab for research purposes.[8] Her mother also developed smallpox, but survived.

The WHO declared smallpox completely eradicated in 1979, the first disease ever thus defeated. Because governments worldwide have ceased vaccination programs, we are faced with huge populations who once again have no native (or in the case of vaccinated individuals, acquired) immunity to his disease. Recently, fears of biological terrorism have led to a call for the reintroduction of smallpox vaccinations. Because there is still no treatment for smallpox, the results would be devastating should the virus get loose in the world.

Smallpox illustrates the fourth factor that can lead to a disease becoming a plague:

- Lack of an effective treatment can lead to uncontrollable spread of a disease.

AIDS

AIDS, or acquired immunodeficiency syndrome, has been recognized as a modern plague. HIV (human immunodeficiency virus) and AIDS are the leading cause of death in Africa today and the fourth leading cause of death worldwide. As a blood-borne and sexually transmitted infection, HIV has disproportionately affected disadvantaged or marginalized people such as commercial sex workers, injection drug users, men who have sex with men (MSM), and people living in poverty. HIV infection has caused approximately 20 million deaths,[9] while 42 million people are estimated to be living with HIV/AIDS.

During 2002 alone, HIV/AIDS caused the deaths of more than 3 million people, including 1.1 million women and 610,000 children under the age of fifteen. The overwhelming majority of people with HIV – approximately 95 per cent of the global total – now live in the developing world.[10] Although AIDS is a viral disease, as yet there is no effective vaccination.

The AIDS epidemic occurred because of all the four factors illustrated earlier.

- It is a new disease for which there was no native (or acquired) immunity.
- At the outset the method of spreading the disease was unknown.
- The very long latency period meant that even once the method of spread was known there were literally hundreds of thousands of infected individuals incubating the disease.
- Until recently the lack of effective (or, in the Third World, affordable) treatment was a major factor in the high death toll due to AIDS.

Variant Creutzfeldt-Jakob Disease (vCJD)

Unknown before 1996, VCJD numbers rose quickly in the United Kingdom. To date, as a result of people eating beef from cattle infected

with mad cow disease, a few hundred cases of vCJD have been diagnosed, but it could well become a plague, with possibly thousands or
even hundreds of thousands of new cases arising in the next ten to
twenty years. The long incubation period of prion diseases suggests that
we've only begun to see the tip of the iceberg that represents vCJD cases.

In order to study prion diseases, scientists have bred certain animals
to be highly susceptible to these diseases. The disease that results from
inoculating these lab animals with prions from cows with BSE is exactly
the same as the disease that results from inoculating them with prions
from a person who died of CJD. It now seems likely that kuru and most
other prion diseases that affect humans are also just variants of the same
disease. All of these diseases consist of a prion infection in the brain. The
initial symptoms probably reflect damage to different parts of the brain
rather than different underlying pathologies.

A 2002 report indicates that some people who appear to have sporadic CJD may in fact be infected with prions that came from cows
infected with BSE.[11] Variant Creutzfeldt-Jakob disease features several of
the factors needed to become a plague.

- It is a new disease for which there is no immunity.
- It has a very long latency period.
- There is no effective treatment.

Alzheimer's Disease

The progression of the Alzheimer's epidemic will have extraordinary
public-health, social, and economic implications. In the United States
alone, it costs at least $100 billion a year to look after people with
Alzheimer's. Furthermore, U.S. businesses lose about $61 billion a year
because of employees who have to care for family members stricken
with Alzheimer's. Other countries face similar hardships.

Alzheimer's disease strikes us as particularly horrible, robbing individuals of those fundamental components of their lives and personalities, perhaps the pieces they cherish most – their memories. AD has been
called a death of a thousand subtractions. Victims lose contact with their
past, become unable to learn new things. As the disease advances, sufferers

become unable to recognize their friends and family; they become fearful, agitated, and, as the disease progresses, unable to care for themselves, getting lost even in their own homes. They lose the ability to speak or have any meaningful communication.

Is it any wonder that elderly people cite fear of this terrible scourge as one of the reasons for contemplating suicide? Many seniors say they fear dementia even more than they fear death. Who would want to continue to live with the knowledge that they are embarking on this ghastly journey where their mind will leave them? Many other seniors worry about being a burden to their families if they get AD.

Alzheimer's kills everyone who gets it. When the victims die, their friends and relatives often breathe a sigh of relief and say things like, "Dad didn't die today – really he died a long time ago. Just his body seemed to carry on." A friend whose mother recently died of AD told us that, even though the loss was very hard, "I lost my real mother years ago."

IS ALZHEIMER'S AN INFECTIOUS DISEASE?

We have already noted the similarity between Alzheimer's disease and CJD and suggested that AD is, in fact, one of the family of prion diseases with which medical science is still beginning to come to terms. Another way of approaching the similarity among these diseases is to focus on the natural history of AD – the circumstances and means by which it is spread.

Characteristics of Infectious Diseases

A plague arises when a population is exposed to a new infectious agent. That means that, as was the case with the Black Death, AIDS, or even vCJD, an epidemiologist (a person who studies patterns of disease spread) can put a date on when the disease first started in a particular population. With AIDS that date was about 1976. With a newly emerging plague, that of SARS (severe acute respiratory syndrome), we can pin down even more accurately the date of onset, at least in North America, to the spring of 2003.

What about Alzheimer's disease? As mentioned earlier, AD appears to have started in about 1900. Prior to that date, we find no evidence of it in any literature, be it mythological, religious, popular, or scientific, and no public-health records to support any evidence for AD.

Medline is the U.S. National Library of Medicine's bibliographic database containing more than 11 million references to journal articles in life sciences such as medicine, nursing, dentistry, and others. We looked at the number of articles that appeared in Medline between 1966 and 2000 for five different conditions: hip fracture, HIV, prostate cancer, Alzheimer's disease, CJD, and myocardial infarction (heart attack). The number of articles cited may be seen as a crude representation of how prevalent these diseases are.

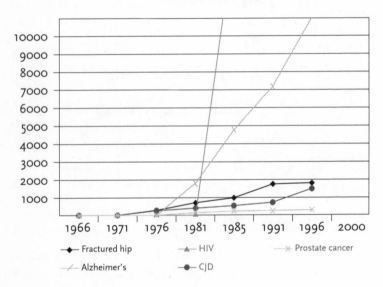

ARTICLES FOUND IN MEDLINE

During that period:

- the number of articles on fractured hips increased by roughly one hundred times;
- articles on CJD increased by fifty times;
- articles on cancer of the prostate increased by six times;
- articles on myocardial infarctions increased by four times

(although the slope of the graph for heart attack was
the same for that as for hip fracture and prostate cancer,
the numbers were ten times larger, so they did not fit on
the graph); and
- articles on Alzheimer's disease increased by *five thousand
times.*

What can account for this incredible disparity? We deliberately chose
five conditions that occur more commonly as people age, so the general
aging of the population cannot account for these differences. New drugs
and surgical procedures have been far more common in the treatment of
cancer and heart disease than in Alzheimer's disease, so new innovations
cannot be the reason.

Interestingly, the increase in the number of articles found for AD con-
forms most closely to HIV – a newly emerged infection – rather than to
the other diseases of the elderly. As the number of cases mushroomed
around the world, so did the number of articles in the literature. One
would expect many more articles on prostate cancer and hip fracture if
the numbers just reflected an aging population, as a multitude of new
treatments and operations for these conditions have come out in the last
twenty years – while there have been no such advances in AD.

The only explanation for these results that makes sense is that
Alzheimer's is far more common now than it was thirty-five years ago.
Once again evidence shows that AD is rapidly changing from a rare condition to a common one. The evidence also reinforces the idea that AD
behaves like an emerging infectious disease – not a chronic disease of
aging.

- The spread of a plague can be documented as travelling
from a point of origin to other areas in the world.

The AIDS epidemic probably started first in Africa, then spread to
North America and Western Europe, and then to less-developed areas
such as China and India. Alzheimer's disease has followed a similar pattern, being first described in Germany, then North America and Western
Europe, and then in Asia. Like all other plagues, AD can be seen to have a

discrete starting point and a well-documented pattern of spread once it became established. So from an epidemiological point of view, AD behaves exactly like an infectious disease.

- An infectious disease gives rise to a pathology similar to other infectious diseases.

This means that all people with a disease caused by bacteria, for example, will have some similar symptoms to other people with bacterial infections. For bacterial diseases, these common symptoms may be a fever or the production of pus. All people with a viral disease produce antibodies to that particular virus.

Patients with prion diseases have a unique set of signs and symptoms. First, they all become demented; second, they all die as a result of the disease; and third, when their brains are examined after death, they all exhibit unique structures called plaques in their brains.

People suffering from AD exhibit exactly these traits. AD is so markedly similar to CJD – the most well known of the prion diseases in humans – that numerous cases in the literature show the two diseases being confused with one another. In some studies, more than 10 per cent of people thought to have died from AD actually died from CJD. Furthermore, because autopsies are rarely performed on people who die of AD, the percentage may be much higher. The opposite has also been found to be true – that is, patients who were believed to have died of CJD turned out to have had AD.[12]

Another striking similarity between CJD and AD is the incidence of the two diseases in any population. While CJD is quite rare, occurring in only about one-in-a-million individuals, it seems to have arisen at about the same time as AD in all countries. For example, AD was first described in 1906 in Germany. The first cases of CJD were described in about 1910 in Germany. Both CJD and AD are first described in Japan in about 1960. Both diseases are still almost non-existent in India.

In the United States, the incidence of CJD is about one in a million at age thirty to forty, and doubles for every five to ten years of age until age sixty to seventy when it appears to level off or drop somewhat.[13]

The incidence of AD in the United States is one in ten at age sixty-five to seventy-five and doubles every decade to age ninety-five.[14]

From a genetic point of view, AD and CJD also exhibit strong similarities. One gene exists that makes a person more susceptible to contracting CJD, and another gene offers a degree of protection from the disease; the same is true with AD. Several studies have shown that the brains of people who have died of these two diseases show biomolecular similarities.[15]

Both AD and CJD arise because of malfunctioning proteins. Studies by Professor Chi Ming Yang from Nankai University in Tianjin, China, suggest that a common molecular mechanism underlies the initiation stages of sporadic Alzheimer's disease and both sporadic and genetic prion diseases.[16] Although AD and prion diseases seem to start in similar ways with the formation of protein plaques, they progress differently. This may explain why Alzheimer's disease advances at a much slower pace than Creutzfeldt-Jakob disease.

- In order to multiply and spread, infectious diseases need reservoirs and vectors.

A *reservoir* is the place where the agent that causes the disease can be found. In many diseases, this can be another species. For example, the reservoir for the Black Death was the rat. For vCJD, the reservoir was cattle that had been infected with prions from eating tainted feed. Sometimes the reservoir can be other humans, as is the case with AIDS and syphilis.

If the reservoir is only human, then when the last case of the disease is eradicated the disease can be said to be eliminated, as was the case with smallpox. Nevertheless, pure specimens of the smallpox virus still exist in at least two laboratories in the world – a situation that has led to concerns about either terrorism or accidents. But barring some disaster precipitated by human evil or stupidity, smallpox can be said to be eradicated.

Monkeypox, a near relative of smallpox, erupted in the United States in early June 2003. Several people got sick after having close contact with imported pet prairie dogs that were sick with monkeypox. This is the first time that there has been an outbreak of monkeypox in the United States.

A *vector* is the method by which the infectious agent is spread from the reservoir to the person who catches the disease. So in the case of the Black Death while the reservoir was rats, the vector was a flea. In the case of syphilis, the reservoir is infected individuals, and the vector or method of spread is sexual contact. In the case of vCJD, the vector is the meat from infected animals. (See table.)

DISEASE	RESERVOIR	VECTOR
Black Death	Rats	Fleas
Syphilis	Infected humans	Sexual contact
Smallpox	Infected humans	Intense personal contact
AIDS	Infected humans	Sexual contact or needle use
vCJD	Infected cattle (from tainted feed)	Meat from infected cattle
West Nile virus	Birds (especially crows)	Mosquitoes

If, in fact, Alzheimer's is an infectious disease and is caused by a prion, how are those prions spread throughout a population? If rats and fleas were the reservoir and vectors for the Plague, what are the reservoir and vector for Alzheimer's disease? Now that rats and fleas no longer dwell quite so cozily with humans, we do not expect to see another worldwide plague of the Black Death. Because AD is such a modern disease, we look for an answer to its origin in our modern lifestyle. The emergence of vCJD caused by eating infected beef gives us our biggest clue to where to look for the source of prion diseases.

HOW DO YOU GET RID OF A PLAGUE?

To keep a plague going, you need a source or cause of infection (reservoir) and a means of disseminating or spreading it (vector). To stop a plague, you can interrupt it at either of these points.

Although certain countries experience regular outbreaks of hemor-rhagic fevers such as Ebola or Lassa fever, and even though these dis-eases are extremely contagious, it's unlikely we will ever have a plague of these diseases – they're too fast. When a person gets one of these dis-eases, she or he dies within a week or two. The outbreak subsides quickly because the reservoir is kept in check.

To get rid of malaria or West Nile virus, you must get rid of the mos-quitoes that are the vectors. However, the means of getting rid of the mosquitoes usually creates controversy. Which is more dangerous, the threat of a mosquito-borne illness, or the harm caused by rampant use of pesticides?

In the past, the pesticide DDT was used extensively against mosqui-toes, to the point where it is now found in the tissues of every person on earth. Forty years after environmentalist Rachel Carson exposed the dan-gers of DDT in her book *Silent Spring*, and thirty years after the United States banned DDT, its chemical fingerprint still shows up in the bodies of American adolescents, according to the Centers for Disease Control. As early as 1960s, biologists had linked DDT to the disappearance of songbirds and raptors. At the same time that its effectiveness has waned against malaria, DDT has been indicted for causing cancer in humans. Its use has since been banned in many countries because of its long-term persistence in the environment. However, because DDT is cheap, some countries continue to manufacture large quantities of it, and farmers spray it on crops. (In what has been called "salad roulette," North Americans often end up ingesting the sprayed produce when they buy imported fruits and vegetables.)

Influenza has often been difficult to control because of rapidly mutat-ing viruses in the reservoirs. The "Asian flu" and "Hong Kong flu," pan-demics of 1956–1957 and 1967–1968, killed a combined 4.5 million people. In those cases, the reservoirs resided in chickens and ducks. In 1997, Hong Kong authorities slaughtered more than a million chickens because of widespread "bird flu" (that came to be known by its genetic nickname of H5N1) among commercial poultry.

In spring 2004, more than 50 million birds were slaughtered in Asia because of an outbreak of bird flu. Although the virus usually causes relatively mild disease in humans, virologists are concerned that if the bird flu virus continues to infect people, it may get a chance to swap genes with ordinary human flu viruses, which could result in a new virus. This new virus could be sufficiently different from the human form that we will have no immunity to it. Scientists fear that such a virus could spark a lethal pandemic.[17] One of the theories of the origin of the 1918 Spanish flu that killed more than 25 million people worldwide suggests just such an occurrence.

In a similar attempt to control infection, authorities in the United Kingdom ordered millions of cattle destroyed because of mad cow disease (BSE), an infectious disease caused by prions that was found to be transmissible to humans in the form of vCJD. Has the reservoir of BSE been wiped out entirely in the U.K.? Only time will tell. Will we see more victims of vCJD? Almost certainly. Because of the long incubation period of vCJD, we will not know for many more years how many victims will develop the disease as a result of eating infected meat. Estimates range from hundreds to hundreds of thousands.

If Alzheimer's disease is in the same way caused by prions, how could we eliminate the source of this infection? Where in the cycle could we interrupt the epidemic? Do modern meat-packing methods contribute to the spread of prion diseases? Certainly that proved to be the case with vCJD. Can eating hamburgers or sausages give you Alzheimer's disease? Would it be possible to eliminate completely the plague of AD? Could the word *senile* ever again come to mean simply old, and not old and demented?

CHAPTER 16

MEAT PACKING A PUNCH

So far, over the course of this book, we have been looking at patterns within and between some disease-related phenomena: varieties of dementia, its incidence and cause, the similarities among a family of neurological diseases including kuru, CJD, and BSE; the discovery and significance of prions; and the rise of the modern meat-packing industry – apparently coinciding with the emergence of some prion diseases. We know that there is a direct relationship between eating BSE-infected meat and vCJD, just as there is a connection between eating the flesh of the victim of kuru and the spread of that disease. In the following chapters, we will see to what extent this established link gives a clue to the spread of other prion diseases – including Alzheimer's disease.

"AN EXTREMELY MINUTE POSSIBILITY"

In August 2002, a Canadian man, under the age of fifty, died of new variant Creutzfeldt-Jakob disease in the western province of Saskatchewan. The man had undergone a surgical procedure prior to his death, and the instrument, an endoscope, that had been used on him was subsequently used in seventy-one other patients. An endoscope is a fibre-optic hose-like device used for internal examinations in a person's throat or rectum.

Normal disinfection and cleaning, which kills all bacteria and viruses, does not kill the malformed proteins that cause prion diseases such as vCJD.

Numerous levels of officials immediately hastened to reassure the public.

- Stephen Whitehead, deputy medical health officer with Regional Health Authority No. 6, said there was "an extremely minute possibility of risk" that there could have been some contamination of the endoscope. (Officials notified the seventy-one other patients that there was a "remote chance" some medical equipment used on them may have been exposed to vCJD, and that they face a "minute risk" of contracting the disease.)
- Ross Findlater, chief medical health officer for Saskatchewan: "There is no risk to the general public from this case."
- Health officials said they did not suspect there were any other cases of the disease in Canada. Antonio Giulivi, Health Canada: "This is the only one."
- Antonio Giulivi: "There is no evidence that mad cow disease has entered the Canadian food supply, and therefore we can reassure the Canadian public that the person did not contract the disease in Canada."

Despite the fact that there have been only a small number of reported cases worldwide of instrument transmission of classical CJD (via surgical instruments used on infected brain tissue), and no transmission of vCJD, we really have no way of knowing how many cases have never been traced. The extremely long incubation period of prion diseases, coupled with the fact that CJD and Alzheimer's disease are often misdiagnosed for each other, means that numerous cases could be slipping through the cracks.

In 1998, Health Canada launched a national CJD surveillance system to monitor for both the classic form of Creutzfeldt-Jakob disease – which strikes about thirty Canadians a year (similar to the worldwide rate of about one case per million population) – as well as for vCJD in Canada. Canadian neurologists, geriatricians, neurosurgeons, neuropathologists, infection-control practitioners, and infectious-disease physicians report all suspected cases of classical or VCJD to a toll-free line, where a standard questionnaire is used to collect information. Health Canada routinely investigates about eighty to one hundred reports of suspected classical CJD and a few reports of suspected VCJD annually.

The CJD Surveillance Unit (CJDSU) in the United Kingdom, set up in 1990 during the BSE crisis, continues at the Western General Hospital in Edinburgh. The main objective of the CJDSU is to identify all cases of CJD in the U.K., and to investigate each case further by clinical examination, clinical investigations, neuropathological examination, genetic analysis, and molecular biological studies, collecting basic epidemiological data and carrying out a case-control study. The unit monitors CJD in order to identify any change in the pattern of this disease that might be attributable to the emergence of BSE.

The decentralized U.S. system, by way of contrast, results in little communication across the country, thus making it difficult to track trends. The National Prion Disease Pathology Surveillance Center was established at the Division of Neuropathology of Case Western Reserve University in September 1997. But except in a few states, CJD is not a reportable disease. CJD surveillance in the U.S. is a somewhat random process, having no official standard questionnaire to help in diagnosis. Those who die with dementia may or may not be autopsied, so the statistics on the rates of CJD, Alzheimer's disease, and other neurological diseases may be seriously flawed.

Notwithstanding the well-regarded Canadian surveillance system, the 2002 VCJD death in Saskatchewan should be cause for apprehension – if not alarm – everywhere.

- All authorities agreed that the man must have contracted VCJD from eating beef during his lengthy stays in Britain because there was no BSE in Canada. But a year and a half after his death, mad cow disease (BSE) was indeed found in Canada – in a cow born and raised in Canada. Because of the long incubation period of prion diseases, it is impossible to tell exactly when and where he contracted the disease.

- As a precautionary measure, the hospital advised those individuals exposed to the possibly contaminated endoscope not to donate blood, organs, or tissues. Furthermore, if any of them had donated blood since their procedure, their blood components that had not been pooled were to be retrieved and destroyed. However, they said that if blood components donated by those individuals had already been pooled, "the theoretical and remote risk of CJD transmission does not merit further action."[1] While it is acknowledged that the risk is minimal, to borrow a phrase from the environmental community, "The solution to pollution is not dilution." Either the donated blood is potentially dangerous, or it is not. If there is no danger, why retrieve and destroy the individual blood donations? If there is danger, why not destroy the pooled supply?

- The medical/dental community still has no protocol in place for addressing the issue of instruments used on patients who are later diagnosed with prion diseases. When you have surgery, you have no way of knowing, nor does your surgeon, whether a previous patient (on whom the same instruments were used) has a prion disease that has been incubating for months or years, but which was not recognizable at the time of his or her surgery. The current practice in Canadian hospitals is to destroy the entire set of surgical instruments that were used if the patient that was operated on is diagnosed as having CJD or vCJD. As a set of neurosurgical instruments can cost upwards of $10,000, this is a decision that does not please hospital administrators. Researchers are working on ways to sterilize instruments so that the prion proteins will be destroyed, but as yet there is no universally accepted way to do this.

- If a suspected CJD case turns out, on autopsy, to have been Alzheimer's disease, no further action is warranted. Yet the two diseases are often misdiagnosed for each other. If,

as we believe, they are closely related prion diseases, then
the probability of contaminated instruments could be
thousands of times more likely than currently suspected.

DEADLY, BUT NOT REPORTABLE

Two incongruous facts trouble many American families of CJD victims.
The first is that CJD is considered so deadly that some surgeons have
refused to operate on patients known to be infected. Some hospitals
have refused to admit patients suffering from CJD. Even some *family
members* of CJD victims have been refused treatment by their dental surgeons. CJD is not transmissible through normal human contact with victims. As with AIDS, the danger lies in direct fluid contact or contact with
improperly disinfected medical instruments. When death inevitably
occurs, some pathologists have refused to perform an autopsy on a
deceased CJD patient. Pathologists who do carry out post-mortem examinations wear a mask, goggles, gloves, boots, and a plastic apron.

In many jurisdictions, funeral homes are forbidden to embalm the
bodies of CJD victims, or are advised strongly against it because of the
danger to those handling the corpse. An embalmer who does work with
a CJD body wears three pairs of gloves, an eye/face shield, mask, cap,
jumpsuit, plastic apron, shoe covers, and all other protective garb. The
solid waste is burned and all fluids and organs are considered infectious
material. Even so, regardless of the procedures, the body will not be disinfected. By definition, one cannot technically embalm the body of a
person who has died from CJD. CJD is resistant to formaldehyde and
every other embalming fluid component or disinfecting chemical, including gluteraldehyde, phenol, and alcohol. The disease organism is just as
infectious as it was before the embalming process.

Guidelines for embalmers advise that all instruments, the embalming
table, embalming room, and embalming room personnel will be exposed
to the infectious organism. Even cosmetologists and hairdressers may be
at risk if working with a CJD body. No procedures exist that will guarantee rendering the organism harmless.

The National CJD Surveillance Unit in the United Kingdom advises
against embalming, as do other knowledgeable CJD authorities.

The second fact that troubles many American families of CJD victims is that despite the deadliness of CJD as shown above, the U.S. Centers for Disease Control and Prevention (CDC) has not made CJD a reportable disease. The CDC monitors the trends and current incidence of CJD in the United States by analyzing death certificate information from U.S. multiple cause-of-death data, compiled by the National Center for Health Statistics, CDC. However, autopsies are not typical in suspected cases because the risk of infection is so great, so death certificates may provide a wildly inaccurate picture of the true rate of infection.

CJD is still very rare in India. There have been only twenty to thirty cases ever reported in India, and almost all of these in two cities.[2] Assuming that the disease has been present since 1960 (the year the first cases were reported in Japan and several other non-European or non-North American countries), then less than one case a year has been reported, which amounts to a rough incidence of one in a billion in India. This is about one thousand times less frequent than every other country that has reported the disease.

If the disease first occurred in Europe in about 1920, that would imply that the person who demonstrated symptoms in 1920 must have contracted the prion about ten to twenty years earlier, or about 1910. This calculation is based on cases of CJD caused by either hGH injection or BSE-contaminated meat, which showed a minimum incubation period of ten to twenty years. Similarly, the Japanese cases must have first contracted the prion-tainted material between 1940 and 1950. Using a similar logic, prion-tainted material must still be very rare in India.

Anecdotal evidence suggests that CJD cases listed as "sporadic" may well have been caused by factors that are no longer traceable. Stories abound of people who took bone-meal pills or used bone meal as fertilizer in their gardens, people who ate squirrel or deer meat, people who took human growth hormone supplements, and years later died of CJD. Another tells of a woman diagnosed with Alzheimer's who died in 1996. Only at autopsy was it discovered she had CJD. Before her death she had been taking a "nutritional supplement" containing ingredients such as vacuum-dried bovine brain, bone meal, bovine eye, veal bone, bovine liver powder, bovine adrenal, vacuum dried bovine kidney, and vacuum-dried porcine stomach. Other stories tell of nine CJD victims who lived and died within a fifteen-mile radius of each other.

Building on past cases and outbreaks, in every case of CJD where a cause has been found, that cause was either eating prion-infected meat or being exposed to infected material due to some medical procedure. Of all the cases of CJD described in the medical literature, less than 5 per cent had neurosurgery or received injections.[3] Logic suggests that the rest of the cases might have contracted the disease through eating meat infected with prions.

India encompasses the world's largest vegetarian society. In the early medieval period, the eating of beef became a taboo in India, if only for upper-caste Hindus. Today, 80 per cent of the population of India is Hindu, most of whom do not eat beef (although there are obviously many exceptions to this rule). In only two states – the eastern state of West Bengal and the state of Kerala on the southwestern tip of India – is the slaughtering of cows legal.

Most Hindu meals include only vegetarian food, but even those Indians who are non-vegetarians eat much less meat than people from other countries. Of the Indian families who do eat meat most usually do so only once or twice a week, often on a Sunday afternoon. Many other families may eat meat – often chicken or mutton – only three or four times a year, most often at weddings.

Hindu Brahmins (upper-caste Hindus) and Jains are more strictly vegetarian. However, the more than 75 million Muslims of India (14 per cent of the population) also follow dietary restrictions, such as avoiding pork or pork products. Due to the general taboo against eating beef, many Indians are *de facto* vegetarians – a fact that may explain the low rate of CJD – and Alzheimer's – in India.

What then caused the sudden appearance of CJD in Japan in the 1960s? Japanese dietary habits have changed markedly since before the Second World War. In 1936, Japanese ate 2.2 kilograms of meat per person. The amount more than doubled to 5.2 kilograms in 1960, then rose to 13.4 kilograms in 1970, and 31.2 kilograms in 1995.[4] This amount was wildly out of reach for domestic production, so by 1997 Japan imported more than 94 per cent of its beef from the United States and Australia.[5]

The prion-tainted meat model might explain the cases in Japan and the relative lack of cases in India, but what about Europe and North America? Why did these diseases suddenly begin to appear about 1910 to 1920? Again using ten to twenty years as an incubation time, what happened to the meat-eating habits around the turn of the century?

The late nineteenth century ushered in the Industrial Age. Leaps in technology allowed for the emergence of a relatively prosperous working and middle classes. This increased prosperity was reflected in an increased demand for meat (similar to what happened in Japan in the period from 1950 to 1990). This increased demand coincided with the development of refrigeration. The development of refrigeration in turn spawned entirely new industries, factory cattle rearing and meat processing. And that's where our troubles began.

THE SMOKING GUN?

Until now, medical science has insisted that "sporadic" CJD has been known for many years and was *not* caused by eating mad cow/BSE-infected meat. If the sporadic form of CJD really is a random event – that one-in-a-million chance of bad luck – why has the rate of sporadic CJD in the United Kingdom been rising as quickly as the VCJD form of the disease – which we know is caused by eating infected meat? For the decade prior to November 2002, 588 cases of sporadic CJD were reported in the U.K., compared with fewer than 100 during the 1970s. Recent evidence suggests that the sporadic form may not arise "spontaneously," but could, in fact, be caused by eating meat infected with BSE.

Professor John Collinge, a prion specialist from the Institute of Neurology, University College, London, has shown that, at least in mice, not only VCJD but also the sporadic form of CJD can be transmitted with BSE prions. (Remember that prions are malfolded proteins that induce other proteins to malfold.) In his experiments, Collinge and his colleagues inoculated mice with BSE prions. Some of these mice produced prion molecules with characteristics indistinguishable from that of sporadic CJD in humans. (In those cases, such sporadic CJD prions were produced in addition to the expected VCJD-type prions.) Collinge notes that their research with mice does not definitively prove that the same would happen in humans, but declares that such a conclusion cannot be ruled out.

In a recent article describing their research, Collinge and his colleagues suggest that the same route of transmission, i.e., BSE-infected meat, could explain sporadic (or classical) CJD in humans.

> This finding . . . raises the possibility that some humans infected with BSE prions may develop a clinical disease indistinguishable from classical CJD. . . . In this regard, it is of interest that the reported incidence of sporadic CJD has risen in the U.K. since the 1970s. This has been attributed to improved case ascertainment [the ability to recognize the disease], particularly as much of the rise is reported from elderly patients and similar rises in incidence were reported in other European countries without reported BSE.
>
> However, it is now clear that BSE is present in many European countries, albeit at a much lower incidence than was seen in the U.K. While improved ascertainment is likely to be a major factor in this rise, that some of these additional cases may be related to BSE exposure cannot be ruled out. It is of interest in this regard that a 2-fold increase in the reported incidence of sporadic CJD in 2001 has recently been reported for Switzerland, a country that had the highest incidence of cattle BSE in continental Europe between 1990 and 2002.[6]

Collinge's research suggests that more than one BSE-derived prion strain might infect humans. In other words, it's possible that some victims who appear to have contracted "sporadic" CJD may have a disease arising from BSE exposure, i.e., from eating BSE-infected meat. This would explain why the rate of sporadic CJD in the United Kingdom has been rising as quickly as the vCJD form of the disease.

It's not a great leap to suggest that some cases of "sporadic" CJD in the United States, Canada, and elsewhere could have been the direct result of eating meat from deer, elk, and moose infected with chronic wasting disease (CWD), the equivalent of BSE in wild ungulates (hoofed mammals). Although officials continued to deny that BSE existed in the U.S.

or Canada until last year, they make no such claim about CWD, which is well documented in both countries.

In Europe, a large-scale case-control epidemiological study of sporadic CJD (not vCJD) comprised of 405 CJD patients made just such a connection, linking sporadic CJD to consumption of raw meat and brains. The study also found "a significant increase in risk of [sporadic] CJD . . . for cases exposed to frequent exposure to leather products, and exposure to fertilizer consisting of hoofs and horns."[7]

Add Collinge's new discovery to the European study, and to what we already know about transmitting CJD through the batching of contaminated pituitary glands. From there it's an equally short leap to ponder commercial meat-packing practices and the batching of meat for hamburgers and sausages.

CANNIBAL CONNECTIONS

The Fore eat the infected bodies of their own tribe – clearly cannibalism. They develop kuru. Children are injected with hormones made from infected human pituitary glands – arguably a variation of cannibalism. They develop CJD.

Humans who eat beef infected with mad cow disease develop vCJD. The cannibal connection? Cows, normally herbivorous vegetarians, suddenly in the 1980s, through no choice of their own, had begun to eat beef. Just one infected cow in the mix, and cannibalism in the feedlot equals an epidemic of mad cow disease.

And even though the repercussions of that mad cow epidemic have already proved enormous, they foreshadow a pandemic of far greater proportions and consequence.

Kuru, scrapie, mad cow disease, CJD – all brain-wasting diseases, each of them leading to a hideous progression of events and a certain death. All are classified as transmissible spongiform encephalopathies (TSEs) or, more simply, prion diseases, all caused by that one little protein – followed by another, and another, and another – gone wrong. Even more significant, all are highly transmissible through eating infected meat.

One other disease resembles CJD so much so that the two diseases are often misdiagnosed for each other. It's another disease that was completely unknown before the twentieth century, also a brain-wasting disease that leads to a hideous progression of events and a certain death. Unknown before the rise of meat-packing plants, Alzheimer's disease (AD) is now epidemic in the western world. The link between CJD and AD has not yet been proven, but many scientists today are looking for that link. How both diseases are spread, and the role played by the meat-packing industry, seem to provide compelling clues.

CHAPTER 17

ALZHEIMER'S DISEASE IN DIFFERENT POPULATIONS

In order to test the thesis that Alzheimer's is a prion disease spread through our modern meat-packing methods, we have to look at the geographical distribution of the disease. Where you live seems to play a huge role in determining if you are going to get AD.

Because different countries or regions may use varied diagnostic criteria or methods of assessment, we need to proceed cautiously when comparing estimates of the rates of Alzheimer's in other locations. Diagnosing people with dementia is not difficult in well-educated populations. Many diagnostic tests for dementia rely on written materials or on assumed knowledge of current events or cultural reference points that "everyone knows." The challenge is to distinguish people with dementia from those with low education but no dementia, and further to distinguish dementia from depression. The 10/66 Dementia Research Group, coordinated through the Institute of Psychiatry, King's College, London, has taken up this challenge and developed tests and screening instruments that cope well with the effects of language and cultural diversity, and deal effectively with the problem of educational bias. Their combined methods work equally well in Indian, Chinese, and Latin American centres.

Enough data on rates of Alzheimer's disease now exists to provide a good basis for comparison and analysis. We do know that specific populations within a geographical area may show differences from the area as a whole. For example, AD is rare in North American Cree Indians, while the prevalence is high among African Americans compared to the American population as a whole.

THE NIGERIAN/AMERICAN STUDY – AND ITS STARTLING RESULTS

Why is the rate of Alzheimer's disease greater in some places than in others? If similar populations were identified with significantly lower or higher incidence rates of AD, such information would greatly enhance the search for risk factors that lead to the genesis of AD.

In a study that has become one of most the widely discussed in the last ten years, an international team led by Dr. Hugh C. Hendrie compared the incidence rates of dementia and AD in two groups of subjects with the same ethnic background but widely differing environments. His team screened and assessed a group of more than two thousand Yoruba people living in Ibadan, Nigeria, and a comparable-sized group of African Americans living in Indianapolis, Indiana. The team first conducted a base-line survey (1992–93) and then two subsequent follow-up waves after two years (1994–95) and five years (1997–98). All the subjects were aged sixty-five or older, and did not have dementia at the beginning of the study.

The results were startling. Over a period of several years, these two diverse, elderly, community-dwelling populations with a similar ethnic origin developed greatly different rates of both dementias of all types and Alzheimer's disease in particular. The group living in the United States had more than twice the incidence rate of AD compared with the group living in Africa. Comparing people of the same age, researchers found Alzheimer's disease at these rates:

- Yoruba, 1.15 per cent
- African Americans, 2.52 per cent.[1]

This was the first study, using the same research method at the two sites, to report significant differences in rates of dementia and Alzheimer's disease in two different communities with similar ethnic origins. The thought-provoking underlying implication is that people living in the United States may be exposed to some significant risk factor to which people living in Africa are not exposed.

RATES OF ALZHEIMER'S DISEASE AND OF
PROCESSED MEAT CONSUMPTION

Hamish McRae, associate editor of the *Independent*, and one of Europe's leading futurists, notes that for the last fifty years America's major export to the rest of the world was – and still is – American culture. In *The World in 2020*, his book about the present and future world economic state, McRae observes that everywhere on this planet people are wearing T-shirts, baseball caps, and athletic shoes.[2] American films and pop music can be seen and heard everywhere, and English has become the closest thing there is to a universal language. It seems that every developing country wishes to emulate the United States in dress, pop culture, and language. Along with Nikes, Madonna, and the Academy Awards, the American diet has gained a foothold in the farthest corners of the earth.

According to the American Heart Association, "Rapid acculturation and improvement in economic conditions have led to the disappearance of the protective effects of a healthy diet. Urban dwellers may believe that a diet high in energy and fat, similar to that of Western affluent countries, is a symbol of their new status."[3] As a country becomes wealthier, the diet evolves from predominately grains, beans, and local fruits and vegetables toward meat – and especially beef.

A strong correlation exists between beef consumption, or more correctly the presence of a commercial beef industry, and rates of both Alzheimer's disease and Creutzfeldt-Jakob disease. The parts of India where there is almost no beef industry have almost no CJD and very low rates of AD.[4]

The fact that AD rates are very low in India has not escaped the notice of e-commerce entrepreneurs. Typing in the words *curry* and *Alzheimer's disease* on the Internet search engine Google yields thousands of links to Web pages. Many of these sites are eager to sell you a genuine Indian curry spice mixture that if taken regularly purports to ward off Alzheimer's. (At least one study has shown that the curry spice cur-cumin reduces amyloid protein – characteristic of AD – in laboratory mice.[5] However, no studies have looked at human consumption of curry and rates of AD.)

We also find low rates of AD in isolated communities that get their meat through hunting, such as the native Canadians living on remote reserves. A study of elderly Cree on two reserves in northern Manitoba identified only one case of probable Alzheimer's disease among eight cases of dementia, giving a prevalence of 0.5 per cent for AD and 4.2 per cent for all dementias. A comparable sample of English-speaking resi-dents of Winnipeg, a large city in southern Manitoba, showed a preva-lence of 3.5 per cent for Alzheimer's disease and 4.2 per cent for all dementias.[6] The researchers noted that both groups had identical rates for *all* dementias, which served to point out the significant disparity in rates of Alzheimer's disease between the two groups.

Between 1980 and the 1990s, the rate of AD in China, Japan, and Korea rose significantly. During this same period the amount of beef consumed in these countries showed a similar steep rise. In Korea in 1970, the per capita consumption of beef was 1,188 grams (2.6 pounds). The rate has

KOREAN BEEF CONSUMPTION

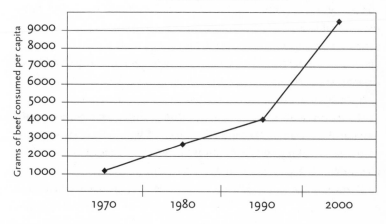

doubled every decade. By 1980, Koreans ate 2,622 grams; in 1990, it had risen to 4,129 grams; and in 2000, Koreans consumed 8,510 grams (18.8 pounds) of beef per person. With the strong tendency toward westerniza-tion of food consumption in recent years, beef has become an integral part of the Korean diet, and Koreans' beef consumption has increased steadily.

Where does all this beef come from? Most of it is now imported. (The United States, Australia, Canada, and New Zealand supplied about 99 per cent of South Korea's imported beef – until May 2003, when Korea and thirty-four other countries banned Canadian beef imports.) Relaxation of beef-importing rules combined with increased consumption has led to increased beef imports. Beef is one of South Korea's largest food imports. In 1970, Korea imported 65 tonnes of beef; ten years later in 1980, the figure had increased by ten times to 6,876 tonnes; in 1990, Korea imported 82,064 tonnes and in 2000 had more than doubled that to 188,124 tonnes.

In the mid-1990s, the Koreans liberalized the trade of processed meats. Koreans now import luncheon meat, canned beef, pork, and ham, and sausage as end-use products. The meat processing industry in South Korea also imports beef for processing.[7]

China and Japan have increased their consumption of beef over the past few decades in a similar fashion. Per capita beef consumption has increased rapidly in China since 1980, but most of China's population traditionally eat pork in preference to beef and mutton, so per capita consumption is still low compared with other developing countries.[8] But rather than relying on imported beef, China is developing their own beef industry – something Japan and Korea cannot do to any extent because of lack of land to raise cattle.

The following charts show the tonnage of meat production and the prevalence of AD in various countries. We have deliberately chosen 1980 as a reference point for meat production because of the long incubation period for prion diseases.

In light of the earlier discussion comparing AD in Nigerians and African Americans, note especially the chart figures for Nigeria, where there is very little commercial meat processing.[9] Remember that preva-lence denotes a snapshot in time – how many people have the disease at this moment (as opposed to how many new cases are arising each year).

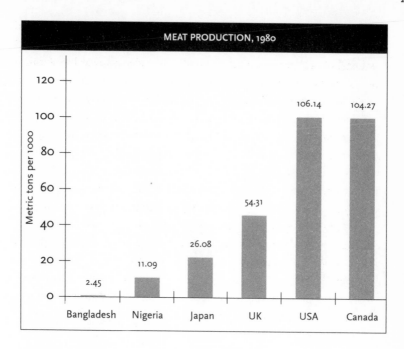

MEAT PRODUCTION, 1980

As the charts show, the rates of Alzheimer's disease parallel reasonably closely the rates of processed meat consumption in these various countries in the 1970s and 1980s – as might be expected if AD is a prion disease, given that incubation periods for prion diseases can be twenty years or more. (Some discrepancies may reflect differing methods of collecting and reporting data, a confusion in diagnosis between CJD, VCJD, and AD, and other confounding factors.)

In China, other Asian countries, and African countries, the rates of AD are consistently shown to be very much lower than the rates in the United States, Canada, the United Kingdom, and other developed countries. As the level of affluence – and consequently the level of meat consumption – rises, the incidence of AD rises in the years thereafter. Prior to the end of the Second World War, both Japan and Korea were essentially non-beef-eating societies, and their rates of AD were very low. After the war, as these countries became more affluent, their rates of beef consumption soared, followed soon by a rise in the rate of AD.

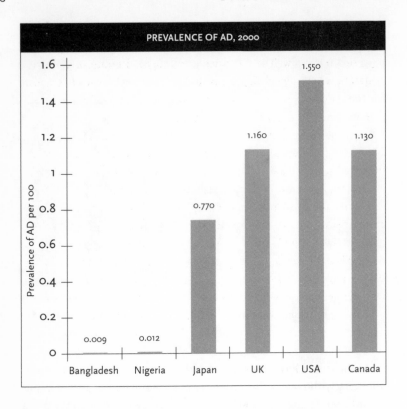

PREVALENCE OF AD, 2000

Reviewing the current and past rates of Alzheimer's disease in various parts of the world in relation to the consumption of processed meat, several facts become evident.

In countries such as India and Bangladesh, where for cultural or religious reasons there is little or no meat-packing or beef importation, the rates of AD are very low. A recent study compared a rural group of seniors in Ballabgarh, India, with a reference U.S. population in the Monongahela Valley of Pennsylvania. In the India group, the overall incidence rate in those aged sixty-five years was 4.7 per 1,000 person-years, substantially lower than the corresponding rate of 17.5 per 1,000 person-years in the Monongahela Valley.[10] This study suggested caution in interpreting their findings, but at the same time noted that these are the first AD incidence rates to be reported from the Indian subcontinent, and they appear to be among the lowest ever reported.

In other countries such as those in northern Africa where for mainly economic reasons the meat-packing industry does not exist, again the rates of AD are low. The low rates in this second group of countries cannot be due to genetic factors, as studies have shown that when native Africans emigrate to the United States, their rates of AD soar even higher than whites – from 14 per cent to as much as 100 per cent higher.[11]

In Canada, the United States, and those countries in Europe where large meat-packing industries have been well established for many years, the rates are high. As the generation that grew up eating fast-food hamburgers begins to age, the rate of AD is rising very quickly.

Exactly the same phenomenon applies with another neurological brain-wasting disease, Creutzfeldt-Jakob. As is the case with Alzheimer's, CJD also is very rare in India, more common in the United States, and is now well established in Japan, where prior to 1960 it was unknown. Many studies note how AD is often confused with CJD and vice versa – and both diseases are on the rise in countries with meat-packing industries.

If, as is postulated, AD is a form of prion disease and may be disseminated in the population by processed meat products, why did it first appear in Germany at the turn of the century? The United States even then had a much larger meat industry than Germany. However, it wasn't until almost twenty years later that cases began to be reported in the United States.

Dietary habits at that time may help to explain timing. In Germany, sausages were, and still are, a major product of the meat-packing industry. Sausages are made in large batches. Sausage makers take meat from several animals, not always even of the same species, mix them together, grind up the mixture, add spices and cereal filler, and put it into casings (made from sheep gut or collagen – an animal by-product). If the meat from even one diseased animal were mixed in such large batches, the infectious particles could spread far and wide.

In the United States at that time, people preferred to eat their beef in large pieces of meat such as steaks or roasts. America's passion for ground beef in the form of hamburgers did not become widespread until the middle of the twentieth century. It was shortly after this time that the current high and rising rates of AD begin to appear in North America.

We've already seen how eating prion-contaminated meat causes vCJD and mad cow disease. The geographical evidence – the pattern of AD rates rising along with the development of modern meat-packing in a given area – strongly suggests that further investigation is urgently needed into the correlation between the consumption of processed beef and the incidence of Alzheimer's disease. It is, of course, difficult to set up experimental controls to account for the many historical and cultural differences among other parts of the world. Specific patterns of meat consumption may not be the only factor that differentiates India from North America, for example, that are relevant to this investigation. But the apparent connection surely merits attention if only because the possible consequences are so dire.

18
CHAPTER

JUST ONE COW?

Canada produces 2.5 per cent of the world's beef supply, most of which has in the past been exported to the United States. In 2000, Canadians consumed an average of 63 kilograms (139 pounds) of red meat per person. In May 2003, Canadians woke up to the news that one cow, a Black Angus from a farm in Alberta in western Canada, had tested positive for mad cow disease (BSE). Within twenty minutes of the announcement, the United States closed its borders to Canadian cattle, and banned importation of all Canadian beef products. In short order, Mexico, Indonesia, Australia, South Korea, Singapore, New Zealand, Taiwan, Philippines, and Japan also banned beef imports from Canada. In total, some thirty-five countries joined the ban. Canadian exports of beef and live cattle to the United States alone exceeded $3.5 billion in 2002, so the ban proved a heavy blow to Canadian cattle farmers, feedlot operators, and numerous related industries.

Guidelines for the minimum number of annual investigations of animals showing characteristics of BSE are provided by the OIE (Office International des Epizooties), the world scientific reference body for animal health. Based on the total adult cattle population, and in accordance with the OIE standard, Canada should test in excess of 300 animals that show neurological dysfunction per year. In 2002, Canada tested a total of 3,377 animals – well above the suggested minimum.[1]

Canadian authorities insisted that the BSE surveillance program was working – they had caught the one cow that had BSE. Furthermore, they had caught it before that cow entered the human food chain. But we may never know if that's true. The head of the cow had been saved – for more than three months – for autopsy, while the rest of the carcass had gone to the renderers, with no one suspecting that the cow had been infected with BSE.

Canadian regulations require rendering plants and feed manufacturers to keep records on the source and distribution of their products. They also specify labelling requirements. For example, the regulations for rendered animal-meat meal require that if the product contains "prohibited material" (i.e., ruminants), it must be labelled with the statement "Do not feed to cattle, sheep, deer or other ruminants."

As part of the investigation into the processing of the infected remains, authorities discovered that some of the remains of the infected cow were turned into poultry feed. Three farms in British Columbia, which had acquired some of the feed, were quarantined because the Canadian Food Inspection Agency (CFIA) could not conclusively determine that ruminant animals on these premises were not inadvertently exposed to the feed. They also found that the rendered material from the cow that tested positive might have been used in the production of dry dog food. Health Canada hastily assured the public that "it is not aware of any available evidence to suggest that physical contact with dry pet food containing meat and bone meal from the infected cow would pose a risk to humans. There is no scientific evidence to date that dogs can contract BSE or any similar disease. In addition, there is no evidence that dogs can transmit the disease to humans." However, it did find it necessary to remind consumers "not to mix dog food into cattle or other animal feeds."[2]

Since August 1997, Canada has banned the feeding of ruminant proteins to cattle, and since January 2001 has banned the import of all rendered animal proteins from countries not designated as BSE-free countries. Given that a team of internationally recognized BSE experts concluded that BSE-contaminated MBM (meat and bone meal feed) probably caused the BSE in Canada's now infamous two mad cows (including the one found in Washington), the system failed on one of two counts (or possibly both).

- Infected material is still being fed to cows, either accidentally or deliberately (illegally),

and/or

- The mad cows ate infected feed before the ban on ruminant proteins.

In either case, the likelihood that these two were the only cows that ate that particular feed and became infected seems exceedingly small. If other cows were also infected, they may have gone to slaughter before they exhibited symptoms of BSE.

Nevertheless, the CFIA declares that BSE controls, including the feed ban, are being followed. The CFIA inspects "a random selection" of Canadian farms each year, along with feed mills and rendering plants throughout the year on a regular basis, and inspection activities are audited annually. Still, they have found it necessary to admit: "When feed mills or rendering plants are found to be out of compliance, the CFIA notifies the facility in writing and conducts follow-up investigations to check that corrective actions have been taken."[3] Clearly some of these facilities have been found to be "out of compliance" on some occasions.

SIX DEGREES OF KEVIN BACON

It's obvious that although several hundred operators are engaged at each stage of the meat-packing process (feeding, slaughtering, rendering, and feed manufacturing), a huge percentage of each of these operations is concentrated in very few places.

The computer and mathematical communities have studied this type of system – known as a scale-free network – in great detail, creating a representative diagram that reveals many individual points or nodes with a few connections, and a few nodes with very many connections (called a hub).

Sociologists tell us that most of us have approximately three hundred acquaintances – people with whom we're on first-name terms. (A quick count in our address book proved that number to be remarkably accurate.) The "small world theory" suggests that everyone in the world can be reached through a short chain of acquaintances. Early work by social psychologist Stanley Milgram suggested that the number of acquaintances between any two people in the world is about six. This claim led to the famous phrase *six degrees of separation*, which in turn led to a play and a movie of the same name. This same pattern of connections is found in many completely unrelated systems such as the Internet or the casts of Hollywood films.

Movie buffs playing "Six Degrees of Kevin Bacon" try to find the links between any actor chosen at random and the actor Kevin Bacon. For example, Will Smith starred in *The Legend of Bagger Vance*, with Charlize Theron; Charlize Theron was in *Trapped* with Kevin Bacon – giving Smith a "Bacon number" of two. Susan Sarandon was in *Last Party* with Courtney Love, who was in *Trapped* with Kevin Bacon, which also gives Sarandon a Bacon number of two. Similarly almost every actor, living or dead, can be linked to Kevin Bacon in two or three steps.[4]

Although hundreds of thousands of actors have appeared in Hollywood movies (in fact, the "Oracle of Bacon at Virginia" Web site uses a database of 575,590 linkable actors), a few such as Kevin Bacon, Rod Steiger, Donald Sutherland, Shelley Winters, and Sean Connery have appeared in many movies (especially movies with large numbers of cast members) and so are hubs that provide links to just about every other actor.

The Internet – essentially a network of networks – also illustrates this phenomenon. Although the Internet consists of well over 4 billion sites, a person can get from any given site to any other site in about nineteen clicks of a mouse on average. This is because sites such as Google and AOL are huge hubs with millions of connections. These networks have come to the attention of scientists because they explain several important biological and social phenomena.

In the biological domain, researchers have found that in cellular metabolism certain proteins act as hubs. That is, a very few proteins are involved in very many cellular reactions while the vast majority are involved in very few reactions. This means that a drug that targets these hub proteins may have very many adverse side effects because that particular protein is involved in so many reactions.

In certain epidemics, the existence or absence of such hubs accounts for the slowness or rapidity of spread of the disease. For example, we can compare the patterns of the spread of lice through a community with the spread of AIDS at its onset.

Almost every child at some point has been exposed to lice through contact with a classmate at school. This infestation can spread to his or her family and other siblings, who can then spread it to their classmates. Each person in this network has roughly the same number of contacts in his or her school, so the infestation spreads in a steady and predictable manner. On the other hand, a single highly promiscuous individual introduced the AIDS epidemic into the North American gay community. Among his hundreds of contacts were a few who also had multiple sexual partners. Because of these few individuals who acted as hubs, the disease rapidly exploded around the world.

The outbreak of severe acute respiratory syndrome (SARS) illustrates the same phenomenon. In May 2003, I spent an uncomfortable week at home in quarantine wearing a mask because I happened to work at a hospital where SARS had occurred. (Fortunately I did not contract the disease, nor did any of my family.) How does SARS spread? A single sick person goes to a hospital and infects a health-care worker. That worker, who comes into contact with numerous others, may serve as a hub, infecting both susceptible individuals and other health-care workers.

These networks have certain characteristics that are important in understanding how to control viruses – of either the Internet or the human variety.

- First, scale-free networks are very resistant to serious damage by random chance. Because the vast majority of nodes have few connections to other nodes, destroying one at random will usually have little effect on the network as a whole. On the other hand, these networks

are very vulnerable to a coordinated attack. Remove (or
infect) one or more hubs, and the entire network will be
severely damaged.

- Second, it is extremely difficult to eradicate completely
  some unwanted feature that can move from node to node.
  In May 2000, the Love Bug, a seriously damaging
  computer virus, shut down major e-mail servers around
  the world, including those belonging to the Pentagon,
  the CIA, the British Parliament, and NASA. Called the
  fastest-moving and most widespread computer virus ever,
  the Love Bug infected systems from Argentina to Japan, to
  New Zealand, and beyond. This virus continued to pop up
  years after its supposed eradication.[5] In 2001, the Code
  Red virus affected about 300,000 computers worldwide,
  programming them to mount a simultaneous attack
  against the White House Web site.

TSE (transmissible spongiform encephalopathy) diseases such as
Creutzfeldt-Jakob disease (CJD) in people and scrapie in sheep are said
to arise spontaneously in one in a million (the "native rate") of the populations known to contract these kinds of diseases. Just as these TSEs do
in all other vulnerable species, similarly, mad cow disease (BSE) must
arise spontaneously in cattle.

Canada has about 13.3 million cattle, the United States about 96.7 million. Even if BSE occurs spontaneously in only one in a million, then at
least thirteen cows in Canada and ninety-seven cows in the U.S. would
naturally have the disease.

Canada slaughters about 3 million cattle annually, the U.S. about 35.4
million. Again, if BSE occurs spontaneously in only one in a million, then
at least three cows in Canada and thirty to forty cattle that are slaughtered annually in the U.S. must have this condition. These admittedly
generalized statistics do not take into account the fact that most cattle
are slaughtered at a very young age, before BSE would be well developed.
What we don't know is *how* infectious the meat would be if the animal
were incubating the disease but had no symptoms whatsoever.

Countries all over the world have reported scrapie in sheep. Scrapie occurs in Canada at a very low level and has been the target of a stringent control program since 1945. Since the first case of scrapie found in a U.S. flock in 1947, this TSE has been diagnosed in more than one thousand sheep flocks in the United States.[6] What would happen if a TSE-infected animal (say a sheep with scrapie) got into the processing system in the United States? If it arrived alive and walking, the meat from that animal would end up in the human food pool. If it were a downer animal, odds are that it would be made into MBM and blood meal that would be used in the manufacture of animal feed (as happened in the United Kingdom). Since 1997, the controls on how animals are rendered and which animals can be used in cattle feed have been in force. However in the years prior to the 1990s, none of these safeguards was in place.

So the chain might look something like this:

> Sheep (possibly scrapie infected) → rendered into MBM →
> fed to pigs and chickens → rendered into MBM → fed to cow
> → human consumption.

Again remember that food intended for pigs and chickens is sometimes mistakenly fed to cattle, so the chain could be shortened to look like this:

> Sheep (possibly scrapie infected) → rendered into MBM →
> fed to cow → human consumption.

CRAZY FOR BURGERS

What happens to beef that is destined for human consumption? After the cow is slaughtered and the inedible portions are sent to the rendering plant, what remains enters the food chain destined for the human population. At the retail level, half of all beef (by poundage) is sold in conventional cuts such as steaks, roasts, and ribs. The remaining half is sold as ground beef. The great majority of ground beef is used in the restaurant industry.

Right from the time of its debut at the 1904 World's Fair in St. Louis, the hamburger proved popular with the public. Almost 20 million people attended the fair, and from that day to this, the popularity of the hamburger skyrocketed. No one knows how many burgers were consumed in St. Louis in 1904, but in 2001 Americans ate 12 billion burgers, 8.2 billion of them in restaurants. Today, burgers account for three-quarters of all beef entrées served in restaurants. Almost 60 per cent of all burgers are purchased in a restaurant.[7]

Another way of looking at this is that on any given day 25 per cent of Americans have a meal at a fast-food restaurant and the vast majority of these meals include a hamburger.

Today, "burgers" might consist of anything from beef or chicken to salmon or even soya, but the classic hamburger, and the one that the most people still eat, is 100 per cent beef. How is this ground beef manufactured? If beef is purchased at a local butcher shop, the butcher may have made it on site from trimmings from other cuts of beef or by grinding lower-cost beef cuts such as chuck. The hamburger purchased at such a shop will probably contain beef from very few animals, as the manufacturing is done using only the small number of animals that a local butcher shop would purchase.

Hamburger meat manufactured for commercial use either in large supermarket chains or in the fast-food industry is another story. In order to ensure a uniform product, beef is mixed with fat in enormous batches. The size of individual batches varies, but a recall of tainted hamburger meat shows the size of potential contamination. Over the course of three days in June 1997, one large processor, Hudson Foods Company of Rogers, Arkansas, produced an average of 404,000 pounds of ground beef per day.[8]

With fat and meat products all coming from different sources, it is impossible to estimate how many individual animals may be represented in a single hamburger patty. The largest cattle plants slaughter steers and heifers to make roasts and steaks, while other plants slaughter cows and bulls. The trimmings from well-fatted steers and heifers are combined with lean cow meat to make ground beef.[9] What happens to the meat and non-meat products derived from a single cow? The diagram below shows how the parts of an individual cow may be distributed. A

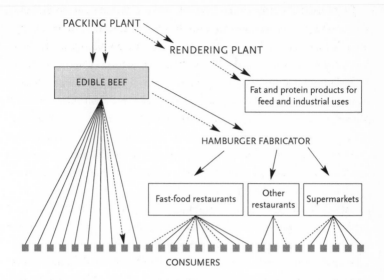

dashed line in this diagram represents a particular single cow. The solid lines represent the vast majority of cows.

The diagram is one of a scale-free network, so once again we can expect any untoward substance that may be found in even a single animal would be widely distributed.

MARJORIE'S DAY

Consumers are not only exposed to beef and its by-products in food. In fact, the penetration of products derived from cattle is so huge the average person may be exposed to these products hundreds of times in a normal day.

Let's look at one day in the life of a North American consumer. For the sake of convenience, let's call her Marjorie.

Marjorie is a lacto-ovo vegetarian, which means she does not eat meat, but she does eat dairy products and eggs. In this sense, she's not the average consumer, most of whom eat meat regularly. She makes an effort to avoid animal products, although she's not fanatical about it. Marjorie might be dismayed to discover how many products from the rendering industry – by-products of the meat industry – are in items she uses each day. Some of them are shown in italics below.

6:30 a.m. Marjorie rises, brushes her teeth with *toothpaste*, rinses with *mouthwash*, and heads for the shower. She uses both *shampoo* and *conditioner* on her hair, applies *shaving cream* to her legs, and washes her face with *soap*. Out of the shower, she splashes on some *skin freshener*, puts on *deodorant*, applies *face cream*, and a light bit of liquid *makeup* and *lipstick*. Putting on her *clothes* and *shoes*, she heads to the kitchen. She pours her coffee in a *bone-china* cup, takes a daily *pill*, and then wolfs down a *muffin* and some *yogurt* for breakfast. She hastily washes the dishes with *detergent*, and throws the dog a *bone-meal biscuit*. She refills the cat's dish with *cat food* and, grabbing her *purse*, heads out the door.

8:00 a.m. Marjorie gets in her car, and settles in to the leather *upholstery*. Just back from the garage, her car has two new *tires* and a new *paint* job, and has been washed and spruced up with car *wax*. She notes that the *asphalt* driveway needs some repair.

2:00 p.m. At work Marjorie flips through some magazines with *high-gloss* covers, and is careful not to get *printing ink* on her hands or on the *textile* upholstery of her chair. The paper in her stationery cupboard has been manufactured with *paper whitener* to make it bright.

5:00 p.m. After work, Marjorie drops off her *photographic film* for processing, and picks up some *fabric softener* and *floor wax* at the grocery store. She fills a prescription for *antibiotics*. To stave off her hunger pangs, she eats a gummi bear *candy*, then pops in a piece of *chewing gum*.

6:00 p.m. Marjorie pours herself a glass of *wine*, while her husband opens a *beer*. The children have a glass of *juice* while they draw pictures with *crayons*, and use *glue* to make craft projects. Lighting the *candles* for atmosphere, they all sit down to dinner – soya burgers, rice, broccoli, and carrots. Marjorie likes *mayonnaise* with her burger, while her husband opts for *margarine* on the bun, along with other condiments. For dessert, they have a choice of *cookies*, *ice cream*, or squares made with *marshmallows*.

7:30 p.m. After dinner, the children toss a *football* around, while Marjorie and her husband work in the garden, applying *fertilizers* and *insecticides*, which they store in their *plywood* and *cement-block* shed.

11:00 p.m. After seeing a television special about mad cow disease, Marjorie puts some *"youth"* cream on her face, and *moisturizer* on her hands, then goes to bed, thankful that because she is a vegetarian, she needn't worry about beef or beef-derived products.

Alas, Marjorie has little idea of the true scope of the rendering industry's encroachment into everyday life. Beef by-products that are sold as edible (for people or animals or both) include blood, hooves, tendons, fat, and all of the animal's organs. The pharmaceutical industry uses all of the animal's glands, including the adrenal, pancreas, thyroid, ovaries, and testes. Most organ meats are sold for human consumption, or they are used in the manufacture of MBM. One exception is the small intestine, which is used in the manufacture of surgical sutures. Although in recent years the use of intestines for sutures and pancreas for insulin has declined, some manufacturers continue to use these products.

Tallow, another by-product of rendering, is used extensively in the manufacture of soaps and detergents as well as in a wide range of lubricants and in the plastics industry. Adhesives are a $154-million industry in the United States, with blood-collagen and blood-soybean adhesive products used in the manufacture of plywood and other fabricated wood products. Tallow-derived products are also used in the manufacture of synthetic rubber. The basic fatty acids that make up the tallow are purified and used in the manufacture of antibiotics.

Gelatin, which is derived principally from hides and bones, is used in large amounts by the food, pharmaceutical, and photographic industries. The food industry uses gelatin mostly in the manufacture of "gummi" type candies as well as many other candy products such as jellybeans, marshmallows, caramels, and fruit chews. In desserts and dairy products, gelatins go into the making of mousse, pie crusts, margarines, yogurts, and ice creams. More esoteric uses in the food industry include clarifying agents for wine making, and gel reinforcement that allows for precooked meats to be sliced more easily.

Most cosmetic products including protective creams, lotions, and shampoos utilize gelatin as a component.

The health and pharmaceutical industry uses gelatins in the manufacture of soft and hard capsules that contain drugs, as well as in biological adhesives and sponges used to control bleeding during surgical operations. The photographic industry uses gelatin extensively in the manufacture of film.[10]

As can be seen from this incomplete list, the average person would be hard-pressed to avoid beef-derived products on a daily basis.

What does all this have to do with the chances of a human coming in contact with some product that was derived from an animal with a prion disease – or specifically from a cow with BSE? This is an impossible question to answer precisely. However, from studying what happened during the BSE outbreak in the United Kingdom, we know certain facts.

- First, although an infected cow is contagious very shortly after being infected, almost none of the infected animals developed overt symptoms of the disease until they were at least three years old or older. As most beef cattle in the United States are slaughtered before they are two, even if the disease were rampant, very few animals would show signs of the disease before being slaughtered.
- Second, with the extraordinary concentration of slaughter, rendering, and feed-manufacturing facilities, even one infected animal would have the potential to contaminate many, many other animals through the rendering process. Up until recently, there have been no restrictions on the practice of using cow and sheep products in the manufacture of animal feed.
- Third, due to the enormous amount of hamburger consumed in both homes and restaurants, the meat from any one particular cow could potentially be eaten by thousands of people. If cattle by-products are factored in, the number of people that could come into contact with some product that was once in a specific cow could reach into the millions. After at least fifty years of batching ground beef both for home and restaurant consumption, and rendering the by-products in huge facilities, it seems inevitable that almost everyone living in any country with this industry would be exposed in some manner to beef products containing prions.

PART SIX

Prevention, Treatment, the Possibility of a Cure

19
CHAPTER

A POPULATION AT RISK

The wretched "lunatics" of Bedlam and the sad, demented patients that I first saw in the back ward as a medical student have more in common with today's AD victims than not. Although we (in most cases) give them better care than ever before, today's unfortunate demented folks will not be cured. Unless something else kills them first, those who suffer from AD will continue to decline mentally until the day they die from the disease.

In 2001, short-term drug therapies for the 4.5 million AD patients in the United States cost an estimated $1.5 billion. These therapies do nothing to cure or prevent the disease, but are aimed almost solely at easing symptoms or delaying the progress of AD. The huge numbers of current cases and the staggering numbers of predicted cases for the next half-century will certainly bankrupt health-care systems worldwide unless a solution is found.

THE NUN STUDY – USE IT OR LOSE IT

If you were looking for a stable population of subjects for medical research, you'd hope for a group with a small number of variables. Because of the uniformity of their lives, nuns living in a convent make ideal subjects for medical research. Unlike most other population groups, they all have the same diet, they all live in the same surroundings, they do roughly the same job, they are all celibate, all receive the same health care, and all have the same income levels.

For this reason, one can safely infer that any changes that may be observed between them that occur over the duration of a study will be due to factors other than diet, environment, occupation, sexual activity, health care, and lifestyle. It would be difficult to find any other group in western society where these variables were the same in such a large group of people for such a long time. Such was the case in the study of aging and Alzheimer's disease that began in 1986 as a pilot study of elderly Roman Catholic nuns, members of the School Sisters of Notre Dame religious congregation in Mankato, Minnesota.[1]

The goal of the Nun Study – which continues to this day – is to determine the causes and means of prevention of Alzheimer's disease, other brain diseases, and the mental and physical disability associated with old age. In 1990, the Nun Study was expanded to include older Notre Dames in other regions of the United States.

The 678 participants in the Nun Study were 75 to 103 years old when the study began, with an average age of 85 years. More than 85 per cent of these participants had been teachers at one time. Participants included women representing a wide range of function and health, from highly functional women in their nineties who have full-time jobs to severely disabled women in their seventies who are bed-bound and unable to communicate. All 678 participants in the Nun Study agreed to donate their brain at death to the University of Kentucky, the home of the Nun Study.

In 1996, epidemiologist David Snowdon, professor of neurology at the University of Kentucky's College of Medicine, published the first of many articles on what was later to be expanded into a book, *Aging with Grace*. This was the first time that "the Nun Study" came into the public awareness.

The most startling result of this study came when the researchers had the opportunity to analyze handwritten autobiographies that many of the nuns had written as young women prior to joining the order. The researchers who analyzed these writings looked for two things:

- Idea density – the number of different or individual ideas expressed for every ten words of text. Idea density is a good measure of a person's language-processing ability.
- Grammatical complexity – the structure of sentences. Grammatical complexity is a good measure of one's working memory capacity.

Initial results showed that these early writing samples offered a good predictor of the women's mental functioning later in life. Low idea density and low grammatical complexity in autobiographies written in early life were associated with low cognitive test scores in late life. This was not too surprising – one would expect that smart children or well-educated children would grow up to be smart adults.

The big surprise came when the nuns died and their brains were analyzed for evidence of Alzheimer's disease. In the initial sample, 90 per cent of the nuns with confirmed (at autopsy) AD had low idea density in the papers they had written some fifty to seventy years earlier. Only 13 per cent among the normal, i.e., non-AD, nuns had low idea density in the autobiographies from their youth. Further research among the same nuns showed that there was about an 80 per cent inverse relationship between idea density early in life and the risk of developing AD. In other words, if a girl had good linguistic ability early in life, she was 80 per cent less likely to develop AD later in life.[2]

These results seem to support the idea, which some have espoused, of the brain as an intellectual muscle. The more you use it, the stronger it gets.

Another study in the same vein found that leisure activities that required intellectual input such as reading, playing a musical instrument, or doing crossword puzzles were associated with a decreased risk of developing AD and vascular dementia. It is unclear whether increased participation in leisure activities lowers the risk of dementia or participation in leisure activities declines during the pre-clinical phase of dementia. Physical activities such as jogging offered no such risk reduction.[3] This study might represent the ultimate revenge of the nerds. Those of us who love crossword puzzles welcomed this report, which I can now use to justify my habit of buying a second newspaper that I really don't read just because it has the best crossword puzzle.

RISK FACTORS FOR ALZHEIMER'S DISEASE

The idea that AD is an infectious disease does not presume that everyone will eventually get it. Just as with other infectious diseases, not everyone who is exposed will develop the disease. Multiple genetic and environmental influences may have an effect on most late-onset cases of Alzheimer's disease. Various studies[4] have shown that certain risk factors make it more likely one will develop Alzheimer's disease. These factors include:

- Old age. Incidence of AD rises steeply with age, at least to age ninety-five.
- Family history of dementia. Having a parent or sibling with AD increases the risk of developing the disease around 3.5 times. The risk is greater for relatives of early-onset cases than for later-onset cases.
- Apo E genotype. Having a particular gene called the apolipoprotein E (apo E) on chromosome 19 affects one's risk of developing Alzheimer's disease. The apo E gene has several different forms, most commonly e2, e3, and e4. A blood test reveals the presence of these forms. Studies indicate that the presence of the e4 form of this gene increases the risk of developing AD, but the risk varies by race. Inheriting two APOE4 genes, one from each parent, increases the risk of AD 33 times in Japanese populations, 15 times in Caucasians, and only 6 times in Africans.[5] Because having the Apo E genotype also affects the risk of developing other neurodegenerative diseases besides AD, some speculate that the gene may play a role in repairing neurons.
- Female gender. Females seem to be at higher risk of AD, but it's uncertain whether this is due to biological or behavioural factors.

- Low levels of education. A low level of education increases the risk of AD significantly, more strongly in women than in men.
- Current smoking increases the risk of AD, more strongly in men than in women.
- Down syndrome. People with Down syndrome invariably begin to accumulate amyloid in the brain – a symptom of AD – before the age of forty. However, they do not generally develop dementia until much older, perhaps suggesting that a buildup of amyloid, although necessary, may not be all that is required for the development of dementia of the Alzheimer's type.
- Geographical location. As we have seen in earlier chapters, where you live affects your risk of developing AD. Countries with a well-established meat-packing industry appear to have a higher rate of AD than countries without such an industry. Remember that Africans in Nigeria have a much lower rate of AD than Africans living in the United States.

Several studies have shown that high blood pressure and high cholesterol are both risk factors for AD, suggesting that vascular factors are also important in Alzheimer's disease, as well as in vascular dementia (MID).

Although some studies suggest that a history of head trauma with unconsciousness is a risk factor for AD, it seems that in fact such a history does not increase risk significantly.

Evidence from the Nun Study showed no statistically significant differences in brain mercury level between Alzheimer's disease and control subjects. Researchers concluded that mercury in dental amalgam restorations does not appear to be a neurotoxic factor in the origin and development of Alzheimer's disease.

20

CHAPTER

THE SEARCH FOR A CURE

The word *prevention*, when used in connection with Alzheimer's disease, suggests that we must prevent people from acquiring the disease – or at least the destructive dementia that accompanies AD. If onset of the disease could be delayed, more people would die of other causes before dying of AD. The best-case scenario would be to prevent anyone from acquiring the disease in the first place.

Effective *treatments* for AD may not stop or reverse the disease but may give temporary respite and relief from some of the debilitating symptoms of the disease.

The word *cure*, when used in connection with Alzheimer's disease, suggests that we must both stop the disease and reverse its effects, an enormous challenge that will require massive funding for research – which becomes more urgent every year as the population ages.

All of the great plagues in history, once established in a population, continued to ravage that population until one of two things happened. Either the entire population, or at least a major portion of it, was eliminated or, as in the case of smallpox, the population acquired immunity to the disease. If there are no susceptible individuals left (those who have been immunized are no longer susceptible individuals), or those that are left are few and scattered, then there is no place for the disease to spread. This most unsatisfactory resolution is what happened when diseases such as smallpox were introduced to the native populations in North and South America after the European conquest. The same thing happened in Europe after the first contact with the Black Death. So many people died, there was no one left to infect. Then, as populations gradually increased again, there would be another outbreak of the plague.

The introduction of foreign plagues and germs has a long and unsavoury role in human history, both real and fictional. In his famous novel *The War of the Worlds*, H.G. Wells used this idea to resolve the plot. The aliens that had invaded Earth from Mars were finally defeated, not by superior technology or force of arms, but by a communicable disease. The invading aliens had no native immunity to the pathogen that rapidly killed every one of them – the common-cold virus.

If present trends continue, this is the method that will evolve with the current plague of Alzheimer's disease. If, as some futurists predict, the average life expectancy by 2050 will be in excess of one hundred years, then based on current rates of AD, at least 80 per cent, and probably more, of those centurions will have the disease. This raises a fundamental question in medical ethics. Is it in the public or individual good to prolong life if it is known that most people who reach old age will have a disease that makes them unable to enjoy their later years and makes them a huge burden on their caregivers, whether family or society?

What can be done about the emerging plague of Alzheimer's disease and the potential plague of vCJD?

The factors that contribute to a plague are:

- a new pathogen for which there is no effective immunity in the native population;
- no effective treatment;
- a long latency period;
- lack of knowledge of, or inability to control, the reservoir of the disease; and
- lack of knowledge of, or inability to control, the vector of the disease.

Prions – malfolded proteins – represent a new pathogen in humans. Kuru, CJD, and other human prion diseases have emerged only within the last century – as has Alzheimer's disease. The way to control the spread of a new pathogen is to increase the level of immunity in the population. There are only two ways that a person can become immune to an infectious disease.

One way to become immune is to have had the disease and recovered. This is a practical method for a disease such as chicken pox, which is a relatively benign disease in childhood but can be very serious in an adult or a pregnant woman. Years ago different neighbourhoods would have the occasional chicken pox party, where all the local young children would be exposed to a child with the disease. Three weeks later all the young guests who had attended the party would come down with the disease, and consequently were guaranteed a lifelong immunity. With diseases with a high mortality, this is obviously not an acceptable method.

The other way to achieve immunity – without having to catch the disease – is to receive a vaccine that will protect the recipient. This is the method that was used so brilliantly in the eradication of smallpox as well as childhood diseases such as measles, mumps, and rubella.

Vaccination is also the goal of those who are working on the eradication of HIV/AIDS. Unfortunately, the first widespread trails of a potential vaccine were not successful, but other vaccines are nearing the human-trial stage. It seems very likely that an effective vaccine will be developed soon. Vaccines are by far the most cost-effective method of dealing with any infectious disease, as they usually require only one or two doses to provide lifetime immunity. For the same reasons, the development of vaccines has been of little interest to most large, multinational drug companies. It makes far more financial sense to develop a drug for a condition such as high blood pressure that will be taken daily for years than to invest money in developing a vaccine that will be taken only once.

Poorer countries such as those in equatorial Africa might be able to afford a national vaccination program for HIV/AIDS, but there is no way they can afford the huge drug costs to treat those already infected.

A VACCINE FOR ALZHEIMER'S DISEASE

Currently there is a great deal of research into a vaccine for AD. In one early study, Dr. Dale Schenk aimed to vaccinate against the amyloid protein that accumulates in the brains of AD sufferers. This study assumed that AD is caused by an accumulation of amyloid – get rid of the amyloid and you will get rid of the disease. Schenk bred a strain of mice that developed amyloid protein in their brains. He then developed a vaccine that allowed the mouse's own immune system to attack and ultimately destroy the amyloid.[1]

This approach worked so well in mice that human trials were instituted. Regrettably, the human trials turned out to be a major disappointment. The vaccine caused serious side effects, including swelling of the brain, which can be fatal. Most disappointing, however, was the fact that even when the vaccine worked, it did not work. The vaccine appeared to be very effective in removing amyloid from the brains of patients with Alzheimer's, but those patients showed no signs of clinical improvement. It was as if someone had airbrushed the X-rays of a patient with lung cancer. The tests were better, but the patient was not. These results seem to indicate that the accumulation of amyloid is a result of the disease, and not the cause.[2]

Other groups are trying to develop different vaccines against other proteins that are present in large amounts in the brains of AD sufferers, although few of these have yet to show any promise.

Every vaccine so far developed in the history of medicine has been aimed at an infectious disease. Although much of the scientific community does not as yet acknowledge that AD may be an infectious disease, they continue to search for a vaccine against it. This fact in itself is perhaps a tacit admission that although the cause of the infection is not yet established, there must be an infectious component to AD. There is currently no research into developing a vaccine for humans against VCJD. A vaccine is effective only if taken before one comes into contact with the disease. Vaccination serves no purpose for people already exposed. In the case of VCJD, because of changes in the meat-packing industry, one hopes that no more people will come into contact with BSE-infected meat.

DRUG TREATMENT FOR AD

The second line of defence against a disease is to develop an effective treatment. Several new drugs have shown some promise in treating both AD and VCJD, and scientists are finding some new uses for older drugs that also show promise.

Currently only two types of treatments have proved to be of any value in the treatment of AD. These treatments fall into the categories of cholinesterase inhibitors and antioxidants.

The first group comprises the drugs known as "smart pills," "brain boosters," or, more formally, cognitive enhancers. These work by decreasing the breakdown of a neurochemical responsible for signalling between neurons in the brain. Drugs that are used to treat depression, such as Prozac, work in a similar fashion. The theory is that levels of these neurochemicals are low in certain diseases. Due to the way that the chemistry of the brain works, these chemicals cannot be given directly, but drugs can be given that slow the natural destruction of these chemicals. Thus the level of these desirable chemicals increases in the brain. In reality what is happening is that those still-functioning brain cells are being driven harder in an attempt to compensate for the cells that have been destroyed by the AD.

Donepezil (brand name Aricept) is the most common drug in this class that also includes galantamine (Reminyl) and rivastigmine (Exelon). The drug memantine has been used for years in Europe but has not yet been approved in Canada or the United States.

How well these drugs work is calculated by how long they delay the progression of the disease. All of them produce essentially the same degree of modest improvement in approximately 30 to 40 per cent of patients with mild to moderate AD.[3] Many well-run studies have shown that patients on these drugs maintain better cognition, as the drugs can delay the progression of disease by between five and nine months.

The second treatment approach works on breaking the cycle of oxidative stress that destroys proteins and perhaps encourages them to aggregate into plaques. Breaking that cycle through antioxidant drugs could be a powerful new therapeutic approach to Alzheimer's disease. Alphatocopherol (vitamin E) belongs to a class of compounds known as antioxidants. Antioxidants may work by slowing the natural death of brain cells. Vitamin E has shown some benefit in some studies, but the gain is quite small, and some studies have shown no benefit at all.

Other drugs being investigated include cholesterol-lowering agents and anti-inflammatory drugs. If any of these prove to be effective, the benefits will probably be small as none of them is targeted at the cause of AD.

Recent studies have disproved any advantage formerly attributed to hormone therapy (a combination of estrogen and progestin) given for women in menopause. Not only does hormone therapy not lower the risk of Alzheimer's disease as previously believed, it actually doubled the risk of AD and other types of dementia in women who began the treatment at age sixty-five or older. Although the risk to an individual woman is slight, the unfavourable finding, along with numerous other drawbacks, led to the great disappointment (and falling profits) of the drug companies who promoted hormone "therapy."

One interesting line of research is looking at preventing proteins from misfolding. In AD, amyloid plaque is deposited in the brain in configurations known as beta sheets, which contain thousands of protein strings. This abnormal folding of brain proteins into sheets seems to cause brain cells to die. Using this fact as a starting point, researchers are trying to develop drugs that will prevent these proteins from misfolding. Compounds called "beta-sheet breakers" that directly target the abnormal structural change appear to hold the best promise for therapies for AD and other neurological diseases such as CJD.[4]

In theory, this is an excellent approach because once the malfolding occurs, the cell dies, so removing the plaque after it has formed will probably be of little benefit, as shown in the failure of the vaccines discussed above. Far better to prevent the misfolding in the first place.

This particular line of research acknowledges that the progression of Alzheimer's disease and prion diseases share common mechanisms. To date no effective drug has been found to prevent proteins from misfolding, but several under current investigation show promise.[5]

TESTING FOR AD

One huge obstacle stands in the way of the development of an effective drug treatment for these or any other prion diseases. That obstacle is the lack of a reliable test to ascertain who has the disease. The only 100 per cent reliable method of determining if someone has AD, CJD, or VCJD is to examine his or her brain under a microscope after death. While this method of testing may be very reliable, it is of no practical use. There is no doubt that X-rays and MRI scans are of value in confirming a diagnosis, but again these tests are expensive and detect the diseases only after the patient is already exhibiting symptoms and the diseases are well established.

For the most part, the diagnosis of AD – and for that matter CJD or VCJD – is a diagnosis of exclusion. If a patient exhibits signs of dementia, that patient is tested for the causes of dementia for which good lab tests exist. Those things that cause dementia for which we can currently test include syphilis, infections in the elderly, and certain vitamin and nutritional deficiencies. If all these tests are negative, then the patient is usually assumed to have AD, or one of the similar dementias such as Pick's disease.

Despite all the advances in medicine and laboratory testing, the diagnosis of AD and CJD is made the same way today as it was when Dr. Alzheimer wrote his paper in 1906 – by talking to and examining the patient. The major flaw in this method, and the reason that a test is so desperately needed, is that by the time the patient begins to show symptoms, the disease is already well established. Furthermore, a test is needed to distinguish between AD and other prion diseases such as CJD. As we have noted, these diseases are often confused.

One of the major factors in the control of syphilis was the development of a test so that patients who were in the latency period (not showing any symptoms) could be treated and cured before the disease did its major damage. In the case of VCJD, we know that many millions of people were definitely exposed to tainted meat in the United Kingdom. But until a test is developed, we do not know whom to treat. Furthermore, even if everyone were treated, how would we know if the treatment is effective? The kuru experience has shown that the disease can occur forty to fifty years after one has been exposed to prion-tainted "meat."

The lack of an early-detection test is of even greater consequence in AD than in VCJD. According to all the incidence and prevalence studies, if they live long enough, at least 80 to 90 per cent of people in North

America and all other countries where Alzheimer's disease is endemic will get the disease. Should an effective method of arresting the progress of the disease become available, these figures are high enough to justify a universal treatment program for everyone over a certain age. But without an effective test, such a program becomes impractical. Cost issues aside, just as with vCJD, how is a researcher to know if the treatment is effective? If treatment were started at, say, age fifty, then with current methods of diagnosis, subjects would have to undergo at least twenty years of medications before it could be ascertained whether the drug worked. These results would be of statistical value only, and would offer no information as to whether any particular patient had benefited from the medications. It would be difficult to convince any regulatory or ethics body to sanction such drug trials.

The search for a test for vCJD and Alzheimer's is not universally supported. Those opposed to a simple test raise two arguments. First, if there is no cure, why would anyone want to know if he or she is going to get it? Unlike AIDS, neither vCJD nor AD can be transmitted from person to person. Those who argued for an AIDS test said, "Even if you might die of the disease, you have a social responsibility to know if you have it so that you will not pass it on to others." This argument does not hold for vCJD or AD – unless it is shown that either can be transmitted through blood products.

The second argument against such a test lies in control of access to the results. If a test were available, especially in places such as the United Kingdom where it is known for a certainty that millions of people were exposed to BSE-infected meat, what would become of the results? Currently there is a debate in the U.K. and the U.S. as to whether results of tests such as DNA analysis should be made available to insurance companies. The patient's point of view argues that even though this information may inform me that I have an elevated risk of developing breast cancer, if I do not currently have the disease, why should I have to pay higher premiums just because I am at risk? Insurance companies argue that if people know they are at risk for some serious disease they will buy large amounts of cheap term insurance, and if they have no significant genetic risks they will not buy insurance at all. If this happened the insurance industry would soon be bankrupt.

This issue is magnified in the case of a test for vCJD. Knowing they would in the future get vCJD, people would wish to purchase large amounts of life insurance; however, if insurance companies know one is at risk of soon developing this disease, they will be reluctant to sell that person any insurance.

This issue is currently being argued in the courts, but early results seem to indicate that in the U.K. people will be obliged to inform the insurance company of any and all test results that they have, including DNA analysis. In the U.S., the opposite is true. The debate continues, and much more will be written on the subject in years to come as more genetic markers for various diseases are identified.

THE HOLY GRAIL – EARLY DETECTION

The holy grail is a test for AD that would detect the disease before any symptoms manifest. Such a test when developed might also be able to detect prion-diseased animals, thereby forestalling the spread of these diseases entirely.

Researchers in many different institutions are working on a blood test that may be able to identify Alzheimer's disease long before symptoms appear. A team from Washington University School of Medicine in St. Louis found that injecting an antibody into mice causes a sudden flood of the amyloid-beta protein in the bloodstream. This is the protein that forms plaques in the brain (a hallmark of Alzheimer's disease). The level of the protein after the injection was an indication of the amount of plaque in the brain. The study in mice found that the blood test could in fact predict the amount of plaque in an animal's brain, and thus could be useful to identify the tendency to form amyloid-beta plaques at an early point in the disease. This method may be useful for quantifying brain amyloid burden in patients who are at risk for AD or in those who have already been diagnosed. Several more years of testing will be needed to find if the test works on humans.[6]

In 2003, Dr. Neil Cashman and his colleagues at the Centre for Research in Neurodegenerative Diseases at the University of Toronto reported early success in finding a possible site where antibodies might bind to prions. This new method relies on the fact that a prion-affected protein has different amino acids "visible" to the immune system. Injecting a specific sequence of amino acids into rabbits, goats, and mice, the researchers found that the animals developed antibodies against it, suggesting they might be immunized against the prion disease. In other words, the antibody recognizes prions. This makes it possible to have a test that will recognize if prion proteins are present in a subject.

The test could be useful in a number of applications, such as in diagnostic testing. Although still in the conceptual stage, it is possible that this discovery could lead to a quick and relatively inexpensive test for those people and animals that have been exposed to a prion disease (such as the millions of Britons who were exposed to BSE-tainted meat). It may even lead to a vaccination for the disease, not only for cattle, but also a treatment for those people in the early stages of the disease. The test could also be useful as a way to monitor the misfolding and distribution of prions in the body.[7]

TREATMENT FOR PRION DISEASES

Although every case of AD, CJD, and VCJD begets tragedy, one particular story provides perhaps the most compelling reason for research into an effective test for these diseases. Jonathan Simms, born in 1984, was a normal, lively, A-level student athlete who was undergoing trials for the Northern Ireland junior international football (soccer) squad in 2001. Healthy and active at six feet, two inches, Jonathan had never smoked and rarely drank, but like the rest of his family, was a regular meat eater. In September 2001, Jonathan's parents first noticed the symptoms of what would soon be diagnosed as variant Creutzfeldt-Jakob disease (VCJD).

The Promise of PPS

As Jonathan quickly spiralled into the hellish world of loss of control and dementia, his frantic father, Don Simms, began searching the Internet and contacting scientists who work in the field of prion diseases. Eventually, Don convinced a physician to try a controversial treatment that involved putting a catheter directly into Jonathan's brain. Using the catheter, the physician delivered the drug pentosan polysulphate (PPS) to Jonathan's brain.

Originally a drug used to treat bladder inflammation in humans, PPS has shown promise in treating prion diseases in animal experiments, and is now being tested as a prophylactic against scrapie in sheep. It is believed that PPS works both to stop the production of new prions and to stop certain kinds of inflammation, although most of its action may be due to the latter effect. It cannot replace dead brain cells. Microbiologist Dr. Stephen Dealler, expert in PPS and spongiform encephalopathies, was instrumental in helping the Simms family to obtain permission for the procedure.

PPS did in fact prove successful in arresting the progression of Jonathan's disease. As of September 2003, he had lived ten months longer than the average vCJD victim. The tragedy is that by the time Jonathan started to receive the drug, his condition had deteriorated to the point that he was in a coma. Although Jonathan's parents did note what seemed to be some new responsiveness in their son, it appears that the damage done by the disease is irreversible. So even if an effective treatment is found, the best that can be hoped for is to arrest the progress of the disease. At the time of writing, approval for PPS treatment has been granted in at least three more cases of patients with prion diseases.

This "solution" is highly unsatisfactory both for AD and vCJD. Both these diseases incubate for years, causing huge amounts of damage to the brain before the first overt symptoms appear. If the disease is arrested, permanent brain damage will destroy any quality of life for both the victim and the victim's family.

In all similar situations where a patient suffers from a terminal illness, the use of heroic measures to prolong life raises serious ethical issues. At what point does further treatment merely prolong the victim's dying, without really enhancing their life?

If PPS should prove to be effective in treating prion diseases, then the necessity for an accurate test becomes even more important. PPS is a drug that requires a surgical procedure merely to administer – a pump has to be placed in the abdominal cavity, and a catheter placed from the pump directly into the brain. This procedure itself carries a significant risk (not only to the patient, but also to the doctors performing the procedure, due to the high infectivity of prion diseases). Although highly dangerous, if this method could be shown to arrest the disease, it could prove to be a lifesaver to those who are currently incubating the disease but not yet showing any symptoms. However, until these patients can be identified, the drug may be capable only of allowing people like Jonathan to remain in that awful purgatory somewhere between living and dead.

A few other drugs have been used with some early promising results, but none has proved to be an effective cure.

Stephen Dealler notes:

> There are now over 40 drugs that are active against prion infection and we have some that are of adequately low toxicity to permit testing on patients.
>
> We expect that some of these drugs and diagnostics will also be useful against Alzheimer's disease, a condition that is extremely difficult to experiment on with animals or in test tubes. So we may well find, in the end, that BSE had a silver lining. We might even end up with a government that does not deny bad news and avoid research in case of what it may find.[8]

Cellular Neuronal Grafting

Research into grafting cells into the brains of prion-disease victims shows that cells actually take on the characteristics of the cell that has been lost or damaged. Currently no cells are available for specific replacement of neurons in prion disease and no studies in live subjects have shown any clinical improvement. But it is possible that within five years some form of treatment using this technique may be possible. This research falls into the category of stem-cell research, which is seriously curtailed in the United States due to its association with embryonic tissues obtained as a "by-product" of therapeutic abortions.

PREVENTING PRION DISEASES

Most current research in AD treatment lies not in prevention or cure, but in slowing the progress of the disease – a promising area of investigation and exploration. Dr. William Molloy and Dr. Paul Caldwell, in their excellent book *Alzheimer's Disease*, recognize the importance of quality of life for AD sufferers.

> Although there is no cure for AD, drugs are now available not only to slow the disease's progress and improve memory, but also to relieve depression, anxiety, and anger. . . . When treating a person with AD it is important to set "goals of treatment". Since there is no cure, the goals of treatment are to improve function and quality of life. It is equally important to decide whether the main goal is to prolong life or improve its quality.[9]

Still, if we could prevent AD and all prion diseases from developing and spreading, the quality of life issue would become next to irrelevant. As with other infectious epidemics, a good method of control for prion diseases is to eliminate the reservoir for the disease. This has largely been done in the United Kingdom for vCJD. Since the strict implementation of the ban against feeding ruminant-derived feed to cattle and sheep, the number of cases of mad cow disease has dropped steadily, and continues to drop. But because cattle and feed were exported prior to the worldwide ban on British cattle and cattle-derived products, BSE is just now emerging in various European and Asian countries. If these countries implement the same control measures that the British did, they too will be free of this disease as soon as their infected cattle have been slaughtered and the cycle renewed.

We have already noted the close connection between the rate of Alzheimer's disease and the consumption of commercial beef, both geographically and temporally. This does not mean that Alzheimer's must be an infectious prion disease like CJD, but rather that prions may play a role in a person's developing the disease.

Dr. Chi Ming Yang, one of the world's leading experts in prion research, outlines a possible mechanism for the pathology seen in Alzheimer's. His research suggests "... a common putative mechanism underlying the initiation stages of sporadic Alzheimer's disease and both sporadic and genetic forms of prion diseases." In other words, a prion may provide the initial push required for the onset of these diseases.[10]

If prions provide the push start for Alzheimer's disease, what can we do to reduce the risk to humans from infectious prions in beef?

The Washington-based Physicians Committee for Responsible Medicine (PCRM) notes the risk of the emergence of mad cow disease in the United States. After reviewing the scientific literature, they conclude:

- The conditions that led to the emergence of BSE in Britain are present in the U.S. Current U.S. livestock rendering and feeding practices are similar to those present in Britain at the onset of the BSE epidemic.
- Evidence suggests that the agent that causes BSE has already spread to at least some animals in the U.S.
- Between 1979 and 1990, 2,614 Americans died of Creutzfeldt-Jakob disease, and the possibility that BSE played a role in some of these deaths cannot be ruled out.

Furthermore, the committee notes:

BSE is not likely to be found until a reasonable number of cattle carcasses are tested for it. While young cattle may harbour disease-causing prions, they are not as likely as older cattle to show pathological brain changes. The increased use of offal in feeds in the U.S. suggests that the risk of BSE is greater now than in the past.[11]

This situation suggests a number of ways to reduce the risk of a prion disease spreading from one cow to another and also from cows to humans.

1. As has been shown in the recent outbreaks of chronic wasting disease in elk, a prion disease can spread from one member of a herd to

another, especially when the animals are kept in close proximity. This would suggest that it might be unwise to confine cows in very close proximity to one another as is currently done in most feedlot operations.

2. All jurisdictions should immediately stop the feeding of any animal-derived protein (including poultry and swine) to cattle and all other non-carnivorous species. There is as yet no demonstrable prion disease of either poultry or swine; however it has been shown in an experimental setting that it is possible to transmit BSE to pigs. Although some animals with a prion disease do not manifest symptoms (perhaps due to their short lifespans), they are capable of spreading the disease to other longer-lived species that eat them. A ban on animal-derived protein feeds would not completely eliminate prion diseases, as it appears that these diseases can arise spontaneously in about one in a million animals. However if such an event should occur, the disease would not be spread to other animals.

3. All jurisdictions should adopt the system of universal precautions in order to prevent the widespread dissemination of infection from the very rare animal that may be infected by a prion disease. This is a method that hospitals implemented a few years ago, and one that has been put into action almost universally since the SARS outbreak. The theory behind this method is that some diseases such as SARS are infectious before the patient actually gets sick. That means that if a patient has been exposed to SARS, he or she can transmit the virus to other people in the days before becoming obviously ill with the disease.

To counter this possibility, and to prevent the spread of similar infectious diseases, all patients are treated as if they are infectious. Health-care workers wash their hands between each patient. Material from every patient such as blood or other bodily fluids are treated and disposed of as if it were infectious. Every person who has been in contact with a person known to have SARS is treated as if they too have the disease, and is isolated until the incubation period has passed. These methods have proved extraordinarily effective in controlling the SARS outbreaks around the world.

Similar methods can be used in the beef industry. The assumption should be that every cow has the potential to be carrying a prion disease, and so the products from that animal should be treated in such a way as to minimize the number of people who may be exposed to this potentially infectious material. If that assumption were universally acknowledged, no animal-derived material would be fed to animals that are used for human food.

As an additional precaution, in most countries improved slaughterhouse procedures are needed to minimize possible contamination. When cattle are stunned with a penetrating bolt, brain material can contaminate other parts of the animal. Stunning methods should only include electrical stunning or non-penetrating captive bolt methods.

4. Specified risk materials (SRM) should be banned from human consumption in all jurisdictions. SRM include skull, brain, nerves attached to the brain, eyes, tonsils, spinal cord, and nerves attached to the spinal cord of cattle aged thirty months or older, and the specified portion of the small intestine of cattle of all ages. Specified risk materials are tissues that, in BSE-infected cattle, have been shown to contain the infective agent and transmit the disease.

High-speed meat-processing equipment used by some meat processors to extract edible tissues from bones can pull tiny amounts of spinal columns and neck bones from cattle into the meat. In 2003, Canada and the United States amended regulations in order to prevent specified risk material (SRM) from entering the human food supply. These materials are also banned from the human food supply in the United Kingdom.

5. Another procedure that would also substantially reduce the potential risk to consumers is to stop the practice of batching that is employed in the manufacture of hamburger, sausages, and other processed meat products. This would mean that the meat that goes into a burger patty would be from only one animal. Such a procedure would not reduce the risk of transmission to zero but would reduce the current risk many thousandfold. If this could be combined with some future effective testing method for prion diseases in food animals, the risk of transmission to people could probably be reduced to almost nothing.

6. The technology already exists for testing cows after slaughter for prion diseases. This is already being done on every cow before it enters the food chain in Japan and in many European countries to a greater or lesser extent. Not surprisingly, the more cows that are tested, the more cases of BSE are being found. The relatively inexpensive test adds about

five cents per pound to the cost of meat. There is no good reason why such a program could not be instituted in North America immediately. The "don't look, don't find" policy that exists now is both wrong and dangerous.

21

CHAPTER

THE FUTURE OF OLD AGE

Greek mythology tells us of Sibyl, a young woman loved by Apollo. When he promised to grant her a wish, she asked to live as many years as she held grains of dust in her hand. Too late she realized that she had forgotten to ask for eternal youth for her long life – so she continued to get older and older for a thousand years, gradually shrivelling up until she could fit inside a tiny bottle. When the children plagued her by asking, "Sibyl, what do you want?" she could merely answer in an endless refrain, "I only want to die."

Like Sybil, we yearn for a long life. But in attempting to increase our longevity, have we forgotten the lesson of Sybil? What will be the fate of the long-lived seniors amongst our citizenry?

The current plague of AD and the potential plague of vCJD represent the greatest public-heath challenge of the twenty-first century. Control of these diseases, and particularly Alzheimer's disease, is pivotal to the entire future of our society.

In some respects we are approaching the limits of the benefits of research into life-prolonging technologies. Is there any real point in eliminating the killers of old age such as heart disease, stroke, and cancer if all the people saved from those diseases will be destined to spend their last years in the mental wasteland of Alzheimer's? If we cannot gain control of the most hideous affliction of old age, perhaps we, like Sybil before us, will find that the treasure of a very long life is really a curse. If this single disease can be eliminated, then the ongoing search for the fountain of youth will have a legitimate and compelling moral justification.

NOTES

NOTES

PART ONE: THE SHORT HISTORY OF ALZHEIMER'S DISEASE

Chapter One: Dementia: Here And There

[1] Lishman, W.A. "Organic Psychiatry." *The Oxford Companion to the Mind.* Richard L. Gregory, ed. Oxford, New York: Oxford University Press, 1989.

[2] Hammond, William A. *A Treatise on Diseases of the Nervous System.* London and New York: Appleton and Co., 1871.

[3] Clouston, T.S. *Clinical Lectures on Mental Disease.* Philadelphia and New York: Lea Brothers & Co., 1898.

[4] Chalmers, John, and Chapman, Neil. "Progress in Reducing the Burden of Stroke." *Clinical and Experimental Pharmacology and Physiology* 28 (12), 1091–1095. December 2001. Online as of March 26, 2003.
<www.blackwell-synergy.com/links/doi/10.1046/j.1440-1681.2001.03582.x/full/>

[5] "What Are the Risk Factors of Stroke?" American Heart Association Inc. 2002. Online as of March 24, 2003.
<www.strokeassociation.org/presenter.jhtml?identifier=1060>

[6] Hein, Hans Ole, et al. "Alcohol consumption, serum low density lipoprotein cholesterol concentration, and risk of ischaemic heart disease: six year follow up in the Copenhagen male study." *British Medical Journal* (1996) 312:736–741. See also: Grønbæk, Morten, et al. "Type of Alcohol Consumed and Mortality from All Causes, Coronary Heart Disease, and Cancer." *Annals of Internal Medicine* Vol. 133, No. 6, (2000).

[7] Walker, Z., and Stevens, T. "Dementia with Lewy Bodies: clinical characteristics and diagnostic criteria." *Journal of Geriatric Psychiatry & Neurology* 15(4):188–94, (2002 Winter). See also: Lennox, G.G., and Lowe, J.S. "Dementia with Lewy Bodies." *Baillieres Clinical Neurology* 6(1):147–166, (April 1977). See also: McKeith, I.G. "Clinical Lewy Body Syndromes." *Annals of the New York Academy of Sciences* 920: 1–8, (2000).

[8] World Health Organization (WHO) statistics on smoking, 2000. Online as of March 11, 2003.
<www.wpro.who.int/public/Regstatistics/reg_info.asp#TOBACCO>
[9] Liu, L.S., "Clinical and Experimental Hypertension: Part A, Theory and Practice." *Hypertension Studies in China* 11(5-6): 859–68 (1989).
[10] Shi, F., Hart, R.G., Sherman, D.G., Tegeler, C.H. "Stroke in the People's Republic of China." *Stroke* 1989; 20. pp. 1581–1585.
[11] American Heart Association. "China CVD Statistics." *Procor, Conference on Cardiovascular Health.* September 5, 2001. Online as of July 30, 2003.
<procor.org/story.asp?storyid=vineetprocor95015&section=S41&sitecode=procor&lang=L1&parentsec=S6&pn=1>
See also: Global Cardiovascular Infobase. <cvdinfobase.ic.gc.ca>
[12] American Heart Association. Heart Disease and Stroke Statistics, 2003 Update. Dallas, Tex.: American Heart Association, 2002.
[13] Suh, G.H., and Shah, A. "A review of the epidemiological transition in dementia: cross-national comparisons of the indices related to Alzheimer's disease and vascular dementia." *Acta Psychiatr Scand* 2001: 104: Issue 1, 4–11. Online as of March 24, 2003.
<www.blackwell-synergy.com/links/doi/10.1034/j.1600-0447.2001.00210.x/full/>

Chapter Two: Dementia: Then and Now

[1] Information on brain development in this section adapted from: Carter, Rita. *Mapping the Mind.* London: Phoenix, 1998.
[2] Bendall, Kate. "This is your life." *Inside Science* 160. *New Scientist* Vol. 178, issue 2395, May 17, 2003.
[3] Rose, Steven. *The Making of Memory.* New York: Anchor Books, 1992. p. 117.
[4] Molloy, Dr. William, and Caldwell, Dr. Paul. *Alzheimer's Disease.* Toronto: Key Porter Books, 1998. pp. 76–78.
See also: Molloy, D.W., Alemayehu, E., et al. "Reliability of a Standardized Mini-Mental State Examination compared with the traditional Mini-Mental State Examination." *American Journal of Psychiatry* 148 (1). January 1991. pp. 102–5.
[5] Molloy and Caldwell, *Alzheimer's Disease.*
[6] Maurer, K., Volk, S., Gerbaldo, H. "Auguste D and Alzheimer's disease." *Lancet* Vol. 349, pp. 1546–1549, 1997.
[7] *Ibid.*
[8] Kraepelin, E. *Lectures on Clinical Psychiatry.* Rev. and edited by Thomas Johnstone. New York: William Wood, 1904.

[9] Facsimile of the 1904 edition. New York, London: Hafner Publishing Company, 1968.

[10] Amaducci, L.A., Rocca, W.A., Schoenberg, B.S. "Origin of the distinction between Alzheimer's disease and senile dementia: how history can clarify nosology." *Neurology*. 36:1497–1499. November 1986.

Chapter Three: Doctors and Medicine in the Old Days

[1] University of Manitoba Medical Catalogue, 1898.

[2] Nordon, Pierre. *Conan Doyle*. Translation of *Sir Arthur Conan Doyle: L'Homme et L'Oeuvre* by Frances Partridge. London: John Murray, 1966.

[3] *Dr. Bell on Sherlock Holmes*. The Sherlock Holmes Society of London. Online as of January 19, 2004. <www.sherlock-holmes.org.uk/The_Society/ Arthur_Conan_Doyle/Joseph_Bell_on_Holmes.htm>

[4] Osler, William, ed. *Modern Medicine: Its Theory and Practice*, Vol. 7. Philadelphia and New York: Lea & Febiger, 1910. p. 711.

[5] Quoted in Bean, William Bennett, M.D., ed. *Sir William Osler: Aphorisms from His Bedside Teachings and Writings Collected by Robert Bennett Bean, M.D. (1874–1944)*. Springfield, Illinois: Charles C. Thomas, 1968.

[6] Osler, William. *Modern Medicine: Its Theory and Practice*. Philadelphia: Lea Brothers & Co., 1935.

[7] Hogan, David B. "Did Osler suffer from "*paranoia antitherapeuticum baltimorenis*"? A comparative content analysis of The Principles and Practice of Medicine and Harrison's Principles of Internal Medicine. 11th Edition." *CMAJ.JAMC* 1999; 161(7):842–845.

[8] Boyd, William. *A Textbook of Pathology: An Introduction to Medicine*. Philadelphia: Lea and Febiger, 1938.

[9] French, Herbert. *An Index of Differential Diagnosis of Main Symptoms*. 62 Walpole Street, London, W.1, March 1912

[10] French, Herbert, ed. *An Index of Differential Diagnosis of Main Symptoms*. London: John Wright and Sons, 1945.

[11] Bouchier, Ian A.D., Ellis, Harold, Fleming, Peter. *French's Index of Differential Diagnosis*. Oxford: Reed Educational and Professional Publishing, 1996.

[12] Roses, Allen D., and Saunders, Ann. "Alzheimer Disease and the Dementias" *Scientific American Medicine*, revised 2002.

[13] Strachey, James, ed., with Anna Freud and Alex Strachey and Alan Tyson. 24 Volumes. London: The Hogarth Press and The Institute of Psycho-Analysis. 1966.

[14] Hunter, Richard, and Macalpine, Ida, eds. *Three Hundred Years of Psychiatry, 1535–1860*. London: Oxford University Press, 1963.

[15] Quoted in Middleton, A.E. *All About Mnemonics*. London: 1885. Online as of February 3, 2003. <pseudonumerology.com/memoes/m1.htm>

[16] Gowers, W.R. *A Manual of Diseases of the Nervous System*. Philadelphia, Pa: P. Blakiston, Son & Co. 1888.

[17] Richardson, Joseph G., et al., eds. *Health and Longevity*. New York: Home Health Society, 1914.

[18] Pierce, R.V., M.D. Buffalo, New York: The World's Dispensary Medical Association, 1935.

Chapter Four: Observations on the Elderly: Ancient Greece to Modern Literature

[1] Finley, M.I. "The Elderly in Classical Antiquity" in *Old Age in Greek and Latin Literature*, Thomas M. Falkner and Judith de Luce, eds. Albany: State University of New York Press, 1989. p. 17.

[2] Minois, Georges. *A History of Old Age: From Antiquity to the Renaissance*. Chicago: University of Chicago Press, 1989. p. 69.

[3] Finley, p. 17.

[4] Haynes, M.S. "The Supposedly Golden Age for the Aged in Ancient Rome: (A study of literary concepts of old age)." *The Gerontologist* 3. 1962. p. 28.

[5] Minois, p. 60.

[6] Aristotle. *Rhetoric*. Translated by Lane Cooper. New York: Appleton, 1932. pp. 134–6.

[7] Bacon, Sir Francis. "History of Life and Death of the Second Title in Natural and Experimental History for the Foundation of Philosophy, Being the Third Part of the Instauratio Magna." 1620.

[8] Gide, A. "Ainsi soit-il." In *Journal, 1939–1949, Souvenirs*. Paris: Gallimard, 1954.

[9] De Beauvoir, Simone. *All Said and Done*. New York: Putnam, 1972.

[10] Skord, Virginia. " 'Withered Blossoms': Aging in Japanese Literature" in *Perceptions of Aging in Literature: A Cross-Cultural Study*, Prisca von Dorotka Bagnell and Patricia Spencer Soper, eds. New York: Greenwood Press, 1989.

[11] Yoshida, Kenko. *Essays in Idleness: The Tsurezuregusa of Kenko*. Translated by Donald Keene. New York: Columbia University Press, 1967.

[12] Tanizaki, Junichiro. *The Diary of a Mad Old Man*. Translated by Howard Hibbett. New York: Knopf, 1965.

[13] Gunn, Edward. "The Honored Aged in Chinese Literature" in *Perceptions of Aging in Literature: A Cross-Cultural Study*, Prisca von Dorotka Bagnell and Patricia Spencer Soper, eds. New York: Greenwood Press, 1989.

[14] "The Twenty-four Examples of Filial Piety (Erh-Shih-szu hsiao)" in *The Four Books*. Translated by J. Legge. Shanghai: Chinese Book Company, 1933.

[15] Minois, Georges. *History of Old Age: From Antiquity to the Renaissance*. Chicago: University of Chicago Press, 1989. p. 305. Minois quotes Harvey C. Lehman, *Age and Achievement*, Princeton, 1953. p. 330.

[16] Stainton, Lilo. "Impressive turnout for workshop on memory." *Daily News* (New York). November 15, 1996.

[17] *Roget's Thesaurus of the English Language*. Garden City, New York: Garden City Books, G.P. Putnam and Son, 1961.

[18] *Roget's International Thesaurus*, New York: Thomas Y. Crowell Company, 1962.

Chapter Five: Age Patterns and Alzheimer's Disease

[1] Hopkins, K. "On the Probable Structure of the Roman Population." *Population Studies 20*. 1966. pp. 245–264.

[2] Demos Lee, Ronald, ed. *Population Patterns in the Past*. London: Academic Press Inc., 1977.

[3] Pyramids based on Troyansky, David G. "The Elderly." *Encyclopedia of European Social History from 1350 to 2000*. Peter N. Stearns, editor-in-chief. New York: Charles Scribner's Sons, 2001. p. 222.

[4] Statistics Canada. Population by Age Group. Online as of July 15, 2003. <www.statcan.ca/english/Pgdb/People/Population/demo31a.htm>

[5] *Encyclopedia of European Social History from 1350 to 2000*. New York: Charles Scribner's Sons, 2001. p. 222. Data from World Health Organization, United Nations, 1998.

[6] *The Decline of Mortality in Europe*, Roger Schofield, David Reher, Alain Bideau, eds. Oxford: Clarendon Press, 1991.

[7] Lancaster, H.O. *Expectations of Life: A Study in the Demography, Statistics, and History of World Mortality*. New York: Springer-Verlag, 1990.

[8] *Encyclopedia of European Social History for 1350–2000*. New York: Charles Scribner and Sons, 2001.

[9] Minois, Georges. *History of Old Age: From Antiquity to the Renaissance*. Chicago: University of Chicago Press, 1989. p. 148.

[10] Alzheimer's Disease and Related Disorders Association, Inc. Online as of July 5, 2002. <www.alz.org/media/understanding/fact/stats.htm>

[11] Molloy, Dr. William, and Caldwell, Dr. Paul. *Alzheimer's Disease*. Toronto: Key Porter Books, 1998.

PART TWO: KURU, CJD, MAD COW, AND VCJD

Chapter Six: Kuru: The Curse of Cannibalism

[1] Milligan, Bruce C. *Milligan's Correlated Criminal Code and Selected Federal Statutes*. St. Catharines, Ontario: Beacon Hill Law Books Ltd., 1995.
[2] Visser, Margaret. *The Rituals of Dinner: The Origins, Evolution, Eccentricities and Meaning of Table Manners*. Toronto: HarperCollins Publishers Ltd., 1991. pp. 8–17.
[3] Sokolov, Raymond. "One Man's Meat Is Another's Person." *Natural History Magazine*. New York: American Museum of Natural History, October 1974, reprinted October 1999. Online as of May 13, 2001.
<www.amnh.org/naturalhistory/index.html?src=h_nh>
[4] Baumbach, Gary, M.D. *Parenchymal Infections: Prions*. Department of Pathology, University of Iowa College of Medicine, 1999. Online as of April 3, 2002.
<www.vh.org/Providers/TeachingFiles/CNSInfDisR2/Text/PInf.CDE.html>
[5] Farquhar, J., Gajdusek, D., eds. *Kuru: early letters and field notes in the collection of D. Carleton Gajdusek*. New York: Raven Press, 1981.

Chapter Seven: Creutzfeldt-Jakob Disease

[1] Quotes taken from anonymous contributors to "*The Many Faces of CJD*." Online as of August 28, 2003.
<www.fortunecity.com/healthclub/cpr/798/cjd.htm>
[2] "Creutzfeldt-Jakob Disease Fact Sheet." Bethesda, MD: National Institute of Neurological Disorders and Stroke. Updated March 15, 2002. Available online as of June 3, 2003.
<www.ninds.nih.gov/health_and_medical/pubs/creutzfeldt-jakob_ disease_fact_sheet.htm>
[3] Prusiner, Stanley B., and McKinley, Michael P., eds. *Prions: Novel infectious Pathogens causing Scrapie and Creutzfeldt-Jakob Disease*. London: Academic Press, Inc. (London) Ltd., 1987.
[4] Manuelidis, Elias E., and Laura Manuelidis. "Suggested Links between Different Types of Dementias: Creutzfeldt-Jakob Disease, Alzheimer Disease, and Retroviral CNS Infections." *Alzheimer Disease and Associated Disorders* 2 (1989): 100–109.
[5] Kemp, Dr. Stephen. *Growth Hormone Deficiency*. eMedicine.com, Inc. January 2, 2002. Online as of October 27, 2002.
<www.emedicine.com/PED/topic1810.htm>

[6] Prusiner and McKinley, eds. *Prions: Novel Infectious Pathogens causing Scrapie and Creutzfeldt-Jakob disease.*

[7] Centers for Disease Control. *Morbidity and Mortality Weekly Report.* June 21, 1985 / 34(24); 359–60, 365–6. Online as of October 22, 2002. <www.cdc.gov/mmwr/preview/mmwrhtml/mm5141a3.htm>

[8] "Senate Community Affairs Legislation Committee Report on The CJD Settlement Offer." The Parliament of Australia. Reviewed August 13, 2001. Online as of October 28, 2002. <www.aph.gov.au/senate/committee/clac_ctte/cjd/ch2_0.htm> See also: Allars, Margaret. "Inquiry into the use of Pituitary Derived Hormones in Australia and Creutzfeldt-Jakob Disease." Canberra, Australia: Australian Government Publishing Services, 1994.

[9] Cooke, Jennifer. *Cannibals, Cows & the CJD Catastrophe: Tracing the Shocking Legacy of a 20th Century Disease.* Sydney: Random House, 1998.

[10] Green, Emily. "A wonder drug that carried the seeds of death" *LA Times*, May 21, 2000.

[11] National Institutes of Health, NIDDK. "Information for People Treated with NHPP Human Growth Hormone (hGH)." April 2003. Available online as of June 3, 2003. <www.niddk.nih.gov/health/endo/pubs/creutz/update.htm#1>

[12] National Institutes of Health, NIDDK. "Information for People Treated with NHPP Human Growth Hormone (hGH)." April 2003. Available online as of August 25, 2003. <www.niddk.nih.gov/health/endo/pubs/creutz/update.htm#1> See also: Johnston, Dr. Lynn, and Conly, Dr. John. "Creutzfeldt-Jakob disease and Infection Control." *The Canadian Journal of Infectious Diseases.* Vol. 12, No. 6. November December 2001. Online as of October 31, 2002. <www.pulsus.com/Infdis/12_06/john_ed.htm>

Chapter Eight: Mad Cows and Dead Englishmen

[1] The BSE Inquiry. Online as of June 12, 2003. <www.bseinquiry.gov.uk>

[2] *Ibid.* <www.bseinquiry.gov.uk/report/volume3/chaptea5.htm>

[3] *Ibid.* <www.bseinquiry.gov.uk/report/volume3/chapterf.htm>

[4] The BSE inquiry. Online as of February 15, 2004. <http://www.bseinquiry.gov.uk/report/volume3/chapter4.htm#516122> See tab 26.

[5] The BSE Inquiry. Online as of June 12, 2003. <http://www.bseinquiry.gov.uk/report/volume3/chaptea5.htm>

[6] *Ibid.* <www.bseinquiry.gov.uk/report/volume3/chaptec6.htm>

[7] The BSE Inquiry. Online as of February 15, 2004.
<http://www.bseinquiry.gov.uk/report/volume3/chapter7.htm>

[8] *Ibid.* <http://www.bseinquiry.gov.uk/report/volume4/chapter9.htm>

[9] The BSE Inquiry. Online as of June 18, 2003.
<www.bseinquiry.gov.uk/report/volume3/chapter4.htm>

[10] The BSE Inquiry. Online as of June 12, 2003.
<www.bseinquiry.gov.uk/report/volume6/chapt412.htm>

[11] *Ibid.* <www.bseinquiry.gov.uk/report/volume8/chapterd.htm>

[12] Sawcer, S.J., et al. "CJD in an individual occupationally exposed to BSE."
Lancet. 341: Mar.6, 1993. p. 642.

[13] The BSE Inquiry. Online as of June 12, 2003.
<www.bseinquiry.gov.uk/report/volume8/chapted2.htm#531779>

[14] The BSE Inquiry. Online as of June 13, 2003.
<www.bseinquiry.gov.uk/report/volume8/chaptec4.htm>

[15] The BSE Inquiry. Online as of February 15, 2004.
<www.bseinquiry.gov.uk/report/volume8/chapter4.htm#559685>

[16] The BSE Inquiry. Online as of June 12, 2003.
<www.bseinquiry.gov.uk/report/volume8/chapted2.htm#531779>

[17] The BSE Inquiry. Online as of June 13, 2003.
<www.bseinquiry.gov.uk/report/volume1/chapte68.htm>

[18] World Health Organization. Variant Creutzfeldt-Jakob Disease: Fact sheet
No. 180. November 2002. Online as of November 18, 2003.
<www.who.int/mediacentre/factsheets/fs180/en/>

[19] Figures from The UK Creutzfeldt-Jakob Disease Surveillance Unit, Western
General Hospital, Edinburgh, Scotland. Online as of November 18, 2003.
<www.cjd.ed.ac.uk/figures.htm>

[20] Dealler, Stephen. "At long last, signs of a BSE breakthrough." *Guardian
Unlimited.* September 5, 2001. Online as of June 16, 2003.
<www.guardian.co.uk/comment/story/0,3604,546972,00.html>

PART THREE: PRION DISEASES

Chapter Nine: Minuscule Assassins

[1] Dobell, Clifford. *Antony Van Leeuwenhoek and His 'Little Animals': Being Some
Account of the Father of Protozoology and Bacteriology and His Multifarious
Discoveries in These Disciplines.* New York: Harcourt, Brace and Company, 1932.

[2] "An abstract of a Letter from Mr. Anthony Leevvenhoeck at Delft, dated Sep. 17,

1683. Containing some Microscopical Observations, about Animals in the scurf of the Teeth." *Philosophical Transactions of the Royal Society of London*. Vol. 14, May 20, 1684, no. 159, pp. 568–574.

[3] Wimmer, Eckard, et al. "Chemical Synthesis of Poliovirus CDNA: Generation of Infectious Virus in the Absence of Natural Template." *Science*. Vol. 297, p. 1016–1018. August 9, 2002.

[4] Comber, Rev. Thomas. *Real Improvements in Agriculture (on the Principles of A. Young Esq.)*. Letters to Reade Peacock, Esq., and to Dr. Hunter, Physician in York, Concerning the Rickets in Sheep. London: Nicholl, 1772. p. 73.

[5] Schwartz, Maxime. *How the Cows Turned Mad*. Berkeley: University of California Press, 2003.

[6] Hadlow, W. "The scrapie-kuru connection: recollections of how it came about." In: Prusiner, S., Collinge, J., et al., eds. *Prion diseases of humans and animals*. Part II. New York: Ellis Harwood, 1992.

[7] Alper, Tikvah, et al. "The exceptionally small size of the scrapie agent." *Biochemical and Biophysical Research Communications* 22 (1966): 278–284.

[8] Alper T., et al., "Does the Agent for Scrapie Replicate without Nucleic Acid?" *Nature* 214 (1967) 764–766.

Chapter Ten: The Ultimate Stealth Invader

[1] "Stanley B. Prusiner – Autobiography." Les Prix Nobel. Stockholm, Sweden: Almqvist & Wiksell International, 1997.

[2] "Science: How Ancient Bacteria handle the Heat." *New Scientist* vol. 131, issue 1781-10, Aug. 1991, p. 21.

[3] Bunk, Steve. "Chaperones to the Rescue." *The Scientist*, Vol. 16, issue 22, 21. Nov. 11, 2001.

[4] Carrell R.W., and Lomas, D.A. "Alpha$_1$-Antitrypsin Deficiency – A Model for Conformational Diseases." *New England Journal of Medicine* vol. 346, no. 1., Jan 3, 2002.

[5] Aguzzi, A., Weissmann, C., "Spongiform encephalopathies: The prion's perplexing persistence." *Nature* 392. April 23, 1998. pp. 763–764.

[6] Nakayama, H., Katayama, K.I., et al. "Cerebral amyloid angiopathy in an aged great spotted woodpecker (Picoides major)." *Neurobiology of Aging* 20(1): 53–56. 1999 Jan-Feb.

[7] Marsh, R.F., and Bessen, R.A. "Epidemiological and experimental studies on transmissible mink encephalopathy." *Developments in Biological Standardization* 80:111–8. 1993.

[8] Zanusso, G., Nardelli, E. et al. "Simultaneous occurrence of spongiform

encephalopathy in a man and his cat in Italy." *The Lancet* Vol. 352, No. 9143. Oct. 3, 1998. pp. 1116–1117.

[9] Dealler, Dr. Stephen. *Lethal Legacy: BSE – the search for the truth*. London: Bloomsbury Publishing Plc. 1996.

[10] Rivera-Milla, E., et al. "An evolutionary basis for scrapie disease: identification of a fish prion mRNA." *Trends in Genetics*. Vol. 19:2: 72–75. 2003.

[11] Bartz, J.C., et al. "Adaptation and selection of prion protein strain conformations following interspecies transmission of transmissible mink encephalopathy." *Journal of Virology*. Vol. 74 (12): 5542–5547. June 2000. Online as of July 4, 2003. <jvi.asm.org/cgi/content/full/74/12/5542>

[12] True, Heather L., and Lindquist, Susan L., "A yeast prion provides a mechanism for genetic variation and phenotypic diversity." *Nature*. 407, 477–483. September 28, 2000.

Chapter Eleven: Other Human Prion Diseases

[1] Brown, P., et al. "FFI Cases from the United States, Australia, and Japan." In *Brain Pathology*, Herbert Budka, ed. Vol. 8:553–570. July 1998.

[2] Tateishi, J., Brown, P., et al. "First experimental transmission of fatal familial insomnia." *Nature* vol. 376:434–5, Aug. 3, 1995.

[3] Figures from the European and Allied Countries Collaborative Study Group of CJD (EUROCJD). Online as of July 7, 2003. <www.eurocjd.ed.ac.uk/allcjd.htm> U.S. figures from National Prion Disease Pathology Surveillance Center. Online as of July 7, 2003. <www.cjdsurveillance.com/report.html>

[4] Kilzer, Lou. "Brain disease's rarity at issue." *Rocky Mountain News*. April 27, 2002. Online as of July 7, 2003. <www.rockymountainnews.com/drmn/state/article/0,1299,DRMN_21_1121130,00.html>

[5] Kimberlin, R.H., Walker, C.A. "Pathogenesis of mouse scrapie; effect of route of inoculation on infectiousness titers and dose-response curves." *Journal of Comparative Pathology*. 1978; 88:39–47.

[6] Prusiner, Stanley B., and McKinley, Michael P., eds. *Prions: Novel Infectious Pathogens Causing Scrapie and Creutzfeldt-Jakob Disease*. London: Academic Press, Inc. (London) Ltd. 1987.

PART FOUR: EATING DANGEROUSLY

Chapter Twelve: How Now Ground Cow

[1] Harris, Marvin. *Cows, Pigs, Wars, and Witches: The Riddles of Culture.* New York: Random House, 1989.
[2] Thomas, Keith. *Man and the Natural World: A History of Modern Sensibility.* New York: Pantheon Books, 1983.
[3] Frink, Maurice, Jackson Turrentine, and Agnes Wright Spring. *When Grass Was King.* Boulder: University of Colorado Press, 1956.
[4] Rifkin, Jeremy. *Beyond Beef: The Rise and Fall of the Cattle Culture.* New York: Dutton, 1992.
[5] Forbes, F.H. "Ice." *Scribner's Monthly Magazine.* Volume 10, num. 4, August 1875. Online as of May 12, 2003.
<cdl.library.cornell.edu/cgi-bin/moa/moa-cgi?notisid=ABP7664-0010-75>
[6] Krasner-Khait, Barbara. "The impact of Refrigeration." *History Magazine* Feb./Mar. 2000.
[7] *Ibid.*

Chapter Thirteen: A Disassembly Plant

[1] Atack, Jeremy, and Passell, Peter. *A New Economic View of American History from Colonial Times to 1940.* New York: W.W. Norton & Company Inc, 1994.
[2] Skaggs, Jimmy M. *Prime Cut: Livestock Raising and Meatpacking in the United States 1607–1983.* Texas A & M University Press, 1986. p. 46.
[3] *Ibid.*
See also: Chicago Historical society Web site
<www.chicagohs.org/history/stock.html>.
[4] Skaggs, p. 109.
[5] Rifkin, Jeremy. *Beyond Beef: The Rise and Fall of the Cattle Culture.* New York: Dutton, 1992.
[6] Maurstad, Lt. Governor David. "Memorandum To: Governor Mike Johanns." January 24, 2000. Online as of May 21, 2003.
<gov.nol.org/Johanns/billrights/ltgovmemo.htm>
[7] Personick, M., and Taylor-Shirley, K. "Profiles in safety and health: occupational risks of meatpacking." *Monthly Labor Review.* Vol. 112, No. 1, January 1989.
[8] House Documents, Vol. 95 65th Congress. 2d Session December 3, 1917–November 21, 1918, Washington: Government Printing Office: 1918. Document No. 1297.

9 Sparks Companies Inc. "The Rendering Industry: Economic Impact of Future Feeding Regulations." Report prepared for The National Renderers Association. June 2001. Online as of May 21, 2003.
<www.renderers.org/economic_impact/sparks.pdf>
10 Pollan, Michael. "Power Steer." *New York Times*. March 31, 2002. Online as of May 22, 2003.
<www.nytimes.com/2002/03/31/magazine/31BEEF.html?ex=1053748800&en=99 bebe5495745f65&ei=5070>
11 Sparks Companies Inc. "The Rendering Industry: Economic Impact of Future Feeding Regulations." Report prepared for The National Renderers Association. June 2001. Online as of May 21, 2003.
<www.renderers.org/economic_impact/sparks.pdf>.
12 Heffernan, William, et al. "Consolidation in the Food and Agriculture System: Report to the National Farmers Union." February 5, 1999. Online as of May 22, 2003.
See also: MacDonald, James M., et al. *Consolidation in U.S. Meatpacking*. Agricultural Economics Report No. 785. USDA/Economic Research Service. March 1999. Chapter 3 "Concentration and Consolidation in Livestock Slaughter." Online as of May 21, 2003.
<www.ers.usda.gov/publications/aer785/>
13 Company information from Tyson Foods, Inc. Web site. Online as of May 22, 2003. <www.tysonfoodsinc.com/company.asp>
14 "*United States Dumping on World Agricultural Markets*." Minneapolis, Minnesota: Institute for Agriculture and Trade Policy. 2002. Online as of May 22, 2003. <www.ictsd.org/issarea/ag/resources/>
15 Pollan, M., "Power Steer."
16 Snapp, Roscoe R. *Beef Cattle: Their Feeding and Management in the Corn Belt States*. New York: John Wiley and Sons, Inc., 1939.
17 Morrison, F.B. *Feeds and Feeding: A Handbook for the Student and Stockman*. Ithaca, New York: The Morrison Publishing Company, 1947.
18 Smith, L.W. "Recycling animal wastes as protein sources." *Alternative sources of protein for animal production, Proceedings of a symposium, National Academy of Sciences, Washington, D.C.*, 1973. p. 146–173.
19 Stanton, T.L., "Feed Composition for Cattle and Sheep." Colorado State University Cooperative Extension, Livestock Series Management No. 1.615, 1999.
See also: Canadian Food Inspection Agency "Animal Feeds." Online as of May 30, 2003. <www.inspection.gc.ca/english/anima/feebet/feebete.shtml>

Chapter Fourteen: Everything but the Moo

[1] David Lamb. Personal correspondence. June 4, 2003.

[2] Dyack, Brenda, and Meilke, Karl D. "Mad Cow Disease and the Value of a Hog in Canada." Canadian Agrifood Trade Research Network CATRN Paper 2001-03. Online as of May 28, 2003.
<www.eru.ulaval.ca/catrn/publications.htmCATRN>

[3] "The Products of Rendering." Rothsay renderers. Online as of May 29, 2003.
<Rothhttp://www3.sympatico.ca/rothsay/products.html>

[4] Canadian Food Inspection Agency Meat Hygiene Directive: Section 4.4.5 – Stunning and Slaughter of Food Animals.

[5] "Forum" hosted by Angie Coiro. KQED Public Broadcasting San Francisco, May 23, 2003.

[6] USDA Animal and Plant Health Inspection Services, Veterinary Services. "Changes in the U.S. Feedlot Industry: 1994–1999." *National Animal Health Monitoring System.* August 2000. Online as of May 27, 2003.
<www.aphis.usda.gov/vs/ceah/cahm/ Beef_Feedlot/fd99chang.htm>

[7] Dyack and Meilke. "Mad Cow Disease..."

[8] The BSE Inquiry. Online as of June 19, 2003.
<www.bseinquiry.gov.uk/report/volume7/execsum4.htm>

[9] Dyack and Meilke. "Mad Cow Disease..."
See also: Meat News. Online as of May 31, 2003.
<www.meatnews.com/index.cfm?fuseaction=Article&artNum=2667>

[10] USDA Animal and Plant Health Inspection Services, Veterinary Services. "Transmissible Mink Encephalopathy." February 2002. Online as of May 27, 2003. <www.aphis.usda.gov/lpa/pubs/fsheet_faq_notice/fs_ahtme.html>

[11] MacKenzie, Debora. "BSE crosses the Atlantic." *New Scientist* vol. 178, issue 2397. May 31, 2003, p. 6.

[12] USDA Animal and Plant Health Inspection Services, Veterinary Services. "Risk Reduction Strategies for Potential BSE Pathways Involving Downer Cattle and Dead Stock of Cattle and Other Species." *Federal Register* (Volume 68, Number 13). January 21, 2002. Online as of May 27, 2003.
<a257.g.akamaitech.net/7/257/2422/14mar20010800/edocket.access.gpo.gov/2003/03-1210.htm>

[13] FDA News. July 11, 2003. Available online as of February 12, 2004.
<http://www.fda.gov/bbs/topics/NEWS/2003/NEW00924.html>

[14] Brown, Paul. "Bovine spongiform encephalopathy and variant Creutzfeldt-Jakob disease." *British Medical Journal.* April 7, 2001; 322:841–844. Online as of August 13, 2003. <bmj.com/cgi/content/full/322/7290/841>

PART FIVE: A MODERN PLAGUE

Chapter Fifteen: Smallpox, Syphilis, AIDS . . . and Alzheimer's?

[1] Haubrich, Dr. William S. *Medical Meanings: A Glossary of Word Origins*. Orlando, Fla.: Harcourt Brace Jovanovich, 1984.
[2] Knox, Dr. Ellis L. "The Black Death." Boise State University. 1995. Online as of May 8, 2002. <history.boisestate.edu/westciv/plague/>
[3] *Ibid.*
[4] Tuchman, Barbara. *A Distant Mirror: The Calamitous 14th Century*. New York: Ballantine Books, 1987.
[5] See "Hospitals in medieval Dalmatia and Dubrovnik." *International Network for the History of Hospitals Newsletter* No. 5, October 2001. Online as of May 8, 2002. <www.cf.ac.uk/hisar/people/kw/newsletter2001.html>
[6] Velendzas, Demetres, MD, and Susan Dufel, MD, FACEP. "Plague." Emedicine.com, Inc. Online as of May 7, 2003. <www.emedicine.com/aaem/topic548.htm>
[7] Gould, Stephen Jay. "Syphilis and the Shepherd of Atlantis." *Natural History Magazine*. New York: American Museum of Natural History, October 2000. Online as of May 13, 2002. <www.findarticles.com/cf_dls/m1134/8_109/65913170/p11/article.jhtml?term>
[8] Kreeger, Karen Young. "Smallpox Extermination Proposal Stirs Scientists." *The Scientist* 8(22):1, Nov. 14, 1994.
[9] Centers for Disease Control and Prevention. "The Global HIV and AIDS Epidemic, 2001." *Morbidity and Mortality Weekly Report*, June 2001. Online as of May 8, 2002. <www.cdc.gov/mmwr/preview/mmwrhtml/mm5021a3.htm>
[10] Joint United Nations Programme on HIV/AIDS as reported by the Centers for Disease Control and Prevention, June 2001. Online as of May 8, 2002. <www.cdc.gov/hiv/stats.htm>
[11] Asante, Emmanuel A., et al. "BSE prions propagate as either variant CJD-like or sporadic CJD-like prion strains in transgenic mice expressing human prion protein." The *EMBO Journal*, Vol. 21, No. 23, pp. 6358–6366, 2002.
[12] Tschampa, H.J., et al. "Patients with Alzheimer's disease and dementia with Lewy bodies mistaken for Creutzfeldt-Jakob disease." *Journal of Neurology, Neurosurgery and Psychiatry*, 2001; 71; 33–39.
See also: Manuelidis, Elias E., and Laura Manuelidis, "Suggested Links between Different Types of Dementias: Creutzfeldt-Jakob Disease, Alzheimer Disease, and Retroviral CNS Infections." *Alzheimer Disease and Associated Disorders* 3 (1989) 100–109.

[13] Holman, Robert C., et al. "Creutzfeldt-Jakob Disease in the United States, 1979–1994: Using National Mortality Data to Assess the Possible Occurrence of Variant Cases." *Emerging Infectious Disease* Vol. 2, Num. 4, Oct-Dec 1996. pp. 333–7. Online as of April 30, 2003.
<www.cdc.gov/ncidod/eid/vol2no4/holman2.htm>

[14] United States General Accounting Office, Report to the Secretary of Health and Human Services, *Alzheimer's Disease Estimates of Prevalence in the United States*, January 1998.

[15] Westaway, D., et al. "Presenilin proteins and the pathogenesis of early-onset familial Alzheimer's disease: beta-amyloid production and parallels to prion diseases." *Prions and Brain Diseases in Animals and Humans: Proceedings of a NATO Advanced Research Workshop, Erice, Italy,* August 19–23, 1996. Morrison, Douglas R.O., (ed.) Vol. 295, p. 159–176; New York, London: Plenum Press. 1998. See also: Yang Chiming (Chi Ming Yang). "An 'i-4, i, i+4' 'reductive and nucleophilic zipper' shared by both prion protein and beta-amyloid peptide sequences supports a common putative molecular mechanism." *Chemical Journal on Internet.* Jul.4, 2000, Vol. 2, No. 7, p. 35. Available online as of April 30, 2003. <www.chemistrymag.org/cji/2000/027035le.htm>

[16] "Link found between proteins responsible for Alzheimer's and vCJD." *The Scientist* August 25, 2000. Online as of May 6, 2003.
<www.biomedcentral.com/news/20000825/07>

[17] Randerson, James, and Debora MacKenzie. "Netherlands' chicken cull tries to stop bird flu virus in its tracks." *New Scientist.* Vol 178, issue 2391. 19 April 2003, p. 10.

Chapter Sixteen: Meat Packing a Punch

[1] Health Canada. "First Canadian Case of Variant Creutzfeldt-Jakob Disease (Variant CJD)." Online as of June 18, 2003.
<www.hc-sc.gc.ca/english/diseases/cjd/index.html>

[2] Satishchandra, P. and Shankar, S.K. "Creutzfeldt-Jakob disease in India (1971–1990)." *Neuroepidemiology.* 1991; 10(1): 27–32.

[3] Rowland, Lewis P., ed., *Merritt's Neurology*, Tenth Edition. Philadelphia, Pa: Lippincott, Williams & Wilkins, 2000.

[4] Ministry of Agriculture, Fisheries & Food, Japan. "Japan's Census Statistics." 1996 Food Charts, 97/98 CD-ROM. p. 64–68. Online as of June 23, 2003.
<www9.ocn.ne.jp/~aslan/2050pfee.htm.>

[5] Statistical Appendix to Agricultural White Paper. Japan, 1998. p. 120. Online as of June 23, 2003. <www9.ocn.ne.jp/~aslan/2050pfee.htm>

[6] Asante, E.A., et al. "BSE prions propagate as either variant CJD-like or sporadic CJD-like prion strains in transgenic mice expressing human prion protein." *The EMBO (European Molecular Biology Organization) Journal*, Vol. 21, No. 23, pp. 6358–6366, 2002.

[7] Van Duijn, C.M., et al., for the European Union (EU) Collaborative Study Group of Creutzfeldt-Jakob disease (CJD). "Case-control study of risk factors of Creutzfeldt-Jakob disease in Europe during 1993–95." *The Lancet*, Volume 351, Number 9109.

11 April 1998, pp. 1081–1085.

Chapter Seventeen: Alzheimer's Disease in Different Populations

[1] Hendrie, H.C., et al. "Incidence of dementia and Alzheimer disease in 2 communities: Yoruba residing in Ibadan, Nigeria, and African Americans residing in Indianapolis, Indiana." *JAMA* 2001 Feb 14;285(6):739–47. See also the earlier report: Hendrie, H.C., et al., "Prevalence of Alzheimer's disease and dementia in two communities: Nigerian Africans and African Americans." *American Journal of Psychiatry* 152:10, October 1995. 1485–1492.

[2] MacRae, Hamish. *The World in 2020: Power, Culture and Prosperity*. London: HarperCollins, 1994.

[3] American Heart Association. *Op. cit.*, 2002.

[4] Chandra, V., et al. "Prevalence of Alzheimer's disease and other dementias in rural India: the Indo-US study Neurology" *Neurology* Vol 51, Issue 4, October 1998. 1000–1008.

[5] Lim, G.P., S.A. Frautschy, and G.M. Cole. "The Curry Spice Curcumin Reduces Oxidative Damage and Amyloid Pathology in an Alzheimer Transgenic Mouse." *Journal of Neuroscience* 21(Nov. 1) 2001. pp. 8370–8377.

[6] Hendrie H.C., et al., "Alzheimer's disease is rare in Cree," *International Psychogeriatrics* 1993 Spring;5(1):5–14

[7] Choi, Jung-Sup, et al., "Beef Consumption, Supply and Trade in Korea." *Agribusiness Review* Vol. 10, 2002. Paper 3 ISSN 1442-6951. Online as of March 14, 2003. <www.agrifood.info/Review/2002v10/Choi/Choi.htm>

[8] Dinghuan, Hu, et al. "Beef Consumption and the Beef Marketing Chain in China." Agricultural and Natural Resource Economics Discussion Paper 4/98. St. Lucia, Australia: School of Natural and Rural Systems Management, University of Queensland. January 1998. Online as of March 14, 2003. <www.nrsm.uq.edu.au/agecon/Pub/pub/discussion/1998/ANREDP98.html>

[9] Chart information from the Food and Agricultural Organization of the United Nations (FAO). All data shown relate to total meat production, that is,

from both commercial and farm slaughter. Data available <www.fao.org> (accessed June 18, 2002).

[10] Chandra, V., et al. "Incidence of Alzheimer's disease in a rural community in India: The Indo-US Study." *Neurology* September 2001; 57:985–989.

[11] Alzheimer's Disease and Related Disorders Association, Inc. "African-Americans and Alzheimer's Disease: The Silent Epidemic." 2002. Online as of March 26, 2003. <www.alz.org/Media/newsreleases/current/021202aareport.html>

Chapter Eighteen: Just One Cow?

[1] CFIA. *Ibid.*

[2] *Ibid.*

[3] Canadian Food Inspection Agency. "BSE Disease Investigation in Western Canada." Online as of November 20, 2003. <www.inspection.gc.ca/english/anima/heasan/disemala/bseesb/bseesbindexe.shtml>

[4] See the Oracle of Bacon at Virginia, University of Virginia. <www.cs.virginia.edu/oracle/>

[5] Buchanan, Mark. *Nexus: Small Worlds and the Groundbreaking Science of Networks.* New York: W.W. Norton & Company, 2002.
See also: Barabási, Albert-László, and Eric Bonabeau, "Scale-Free Networks," *Scientific American* May 2003. pp. 60–69.

[6] USDA Animal and Plant Health Inspection Services, Veterinary Services "Scrapie" Fact Sheet February 2002. Online as of June 1, 2003 <www.aphis.usda.gov/lpa/pubs/fsheet_faq_notice/fs_ahscrapie.html>

[7] "Beef Trivia." Online as of June 2, 2003. <www.iabeef.org/Fun/>

[8] Food Safety and Inspection Service, USDA Release no. 0276.97. Online as of June 2, 2003. <www.fsis.usda.gov/OA/recalls/prelease/pr015-97a.htm>

[9] MacDonald, James M., et al. "Consolidation in U.S. Meatpacking." *Agricultural Economics Report No. 785.* March 1999.

[10] National Cattlemen's Beef Association. "Beef By Product Usage 1996." Online as of June 2, 2003. <www.beef.org>

PART SIX: PREVENTION, TREATMENT, THE POSSIBILITY OF A CURE

Chapter Nineteen: A Population at Risk

[1] See the Nun Study at <www.mc.uky.edu/nunnet/>. Online as of August 22, 2003.

See also: Snowdon, David. *Aging with Grace: What the Nun Study Teaches Us about Leading Longer, Healthier, and More Meaningful Lives.* New York: Bantam Books, 2001.

[2] Snowdon, D.A., et al. "Linguistic ability in early life and cognitive function and Alzheimer's disease in late life. Findings from the Nun Study." *JAMA* Vol. 275 No. 7, February 21, 1996.

[3] Verghese J., et al. "Leisure Activities and the Risk of Dementia in the Elderly." *New England Journal of Medicine,* 2003 June 19;348 (25):2489–90.

[4] Launer, L.J., et al. "Rates and risk factors for dementia and Alzheimer's disease: results from EURODEM pooled analyses." *Neurology.* 1999 Jan 1;52(1):78–84.

[5] Risch Neil, et al. "Categorization of humans in biomedical research: genes, race and disease." Genome Biology.com. Online as of August 22, 2003. <http://genomebiology.com/2002/3/7/comment/2007/abstract>

Chapter Twenty: The Search for a Cure

[1] Schenk, D., et al. "Immunization with amyloid-beta attenuates Alzheimer-disease-like pathology in the PDAPP mouse." *Nature* 400 173–177, 08 July 1999.

[2] Motluk, Alison. "Fragile minds: Could the proposed treatments for Alzheimer's make this debilitating disease even worse?" *New Scientist* vol. 177, issue 2380-01, Feb. 2003, p. 34.

[3] Clark, C.M., et al. "Alzheimer Disease: Current Concepts and Emerging Diagnostic and Therapeutic Strategies." *Annals of Internal Medicine,* 4 March 2003, Volume 138, Number 5, pp. 400–410.

[4] Wisniewski, T., et al. "Therapeutics in Alzheimer's and Prion Diseases." *Biochemical Society Transactions* (2001) 30, (574–578). Online as of August 22, 2003. <www.biochemsoctrans.org/bst/030/bst0300574.htm>

[5] Knopman, D. "Pharmacotherapy for Alzheimer's Disease: 2002." *Clinical Neuropharmacology* Vol. 26, No. 2, pp. 93–101.

[6] DeMattos, Ronald B., et al. "Brain to Plasma Amyloid-beta Efflux: a Measure of Brain Amyloid Burden in a Mouse Model of Alzheimer's Disease." *Science* Volume 295, Number 5563, 22 Mar 2002, pp. 2264–2267.

[7] Paramithiotis, E., Cashman, N., et al. "A prion protein epitope selective for the pathologically misfolded conformation." *Nature Medicine.* July 2003, Volume 9, Number 7; pp. 893–899.

[8] Dealler, Stephen. "At long last, signs of a BSE breakthrough." *The Guardian.* September 5, 2001.

[9] Molloy, Dr. William, and Dr. Paul Caldwell. *Alzheimer's Disease*. Toronto: Key Porter Books, 1998. p. 185.

[10] Yang Chiming. "An 'i-4, i, i+4' 'reductive and nucleophilic zipper' shared by both prion protein and beta-amyloid peptide sequences supports a common putative molecular mechanism." *Chemical Journal on Internet*. July 4, 2000. Vol. 2, No. 7, p. 35. Online as of June 2, 2003.
<www.chemistrymag.org/cji/2000/027035le.htm>

[11] Physicians Committee for Responsible Medicine. "Mad cow disease: the risk to the U.S." 1998. Online as of August 23, 2003.
<www.pcrm.org/health/Preventive_Medicine/mad_cow_disease.html>

INDEX

INDEX

Acquired Immune Deficiency
Syndrome (AIDS). *See* HIV/AIDS
adolescents: brain growth, 23
Africa. *See also specific countries*: AD,
16, 225, 229, 231; dementia, 19;
HIV/AIDS, 203, 207; smallpox, 201;
vaccination versus drug treatment,
254
African Americans, 225, 228
age factors: AD, 250; CJD, 92–93, 115;
DLB, 14; FFI, 151, 154; GSS, 154;
HIV/AIDS, 13; MID, 11, 12
age pyramids, 69
aging. *See* elderly people; literature
on aging
agri-business, 173–75
Alberta, 185, 233
Albertsons (company), 183
alcohol, 13
Alper, Tikvah, 136, 138
Alzheimer, Alois, 33–35
Alzheimer's disease (AD):
and DLB, 15; as modern plague, 196,
204–5, 253, 269; as prion disease,
9–10, 156–57, 209, 264–65; costs,
204, 247; cure, 252; development,
23; diagnosing, 29–33, 224–25,
257–60; early detection, 260–61;
first cases, 33–34, 205, 207–8;
geographical differences, 18–19,
224–32, 251; in the literature, 67,
206–7; link to CJD, 94, 208–9,
216–17, 222–23; link to meat
production and consumption,
226–32, 264; Nun Study, 247–49;
pre-senile and senile, 65; preventing,
252, 265–68; rates of, 15–16, 18–19,
35–36, 54, 76–78, 208–9; recency,
36, 45–48, 50–51, 68, 76, 207, 254;
research, 248–49, 255, 260–61;
risk factors, 250–51; signs and
symptoms, 29–30; treatment, 252,
256–57, 264; vaccine research, 255
amino acids, 138–39
amygdala, 21
amyloid plaques, 141, 255, 257, 260
animal feed: ingredients, 144, 176–78,
180, 188–89; link to TSEs, 110–11,
145, 186–89; made from downer
animal parts, 183, 185, 188, 239;
North American policies, 188–91,
234–35; recommendations to
prevent prion disease, 266
animals: domestic. *See specific animal*:
prion infectivity, 141, 143–44; wild,
93, 145, 149, 190, 221, 265; zoo,
106–8, 146
anomia, 27
antelopes, 106–8, 147
antibiotics, 243
anti-inflammatory agents, 257
antioxidants, 256–57
antiretroviral therapy, 14
Aricept (donepezil), 256

DATE DUE

~~AUG 1 0 2010~~			
OCT 1 9 2011			
~~FEB 2 4 2018~~			